AF798571

PLANT PHYSIOLOGY

Concepts and Practices

PLANT PHYSIOLOGY
Concepts and Practices

Dr. Awani K. Singh

CENTRUM PRESS
NEW DELHI-110002 (INDIA)

CENTRUM PRESS

H.O.: 4360/4, Ansari Road, Daryaganj,
New Delhi-110002 (India)
Tel: 23278000, 23261597, 23255577, 23286875

B.O.: No. 1015, Ist Main Road, BSK IIIrd Stage,
IIIrd Phase, IIIrd Block, Bengaluru-560085 (INDIA)
Tel: 080-41723429

Email: centrumpress@gmail.com
Visit us at: www.centrumpress.com

Plant Physiology: Concepts and Practices

First Edition, 2013

ISBN 978-93-5084-220-1

PRINTED IN INDIA

Printed at AnVi Composers, New Delhi

Contents

Preface

Plant physiology is a subdiscipline of botany concerned with the functioning, or physiology, of plants. Closely related fields include plant morphology , plant ecology, phytochemistry, cell biology, genetics, biophysics and molecular biology.

Fundamental processes such as photosynthesis, respiration, plant nutrition, plant hormone functions, tropisms, nastic movements, photoperiodism, photomorphogenesis, circadian rhythms, environmental stress physiology, seed germination, dormancy and stomata function and transpiration, both parts of plant water relations, are studied by plant physiologists. The field of plant physiology includes the study of all the internal activities of plants—those chemical and physical processes associated with life as they occur in plants. This includes study at many levels of scale of size and time. At the smallest scale are molecular interactions of photosynthesis and internal diffusion of water, minerals, and nutrients. At the largest scale are the processes of plant development, seasonality, dormancy, and reproductive control. Major subdisciplines of plant physiology include phytochemistry and phytopathology. The scope of plant physiology as a discipline may be divided into several major areas of research. Plant physiology includes the study of biological and chemical processes of individual plant cells. Plant cells have a number of features that distinguish them from cells of animals, and which lead to major differences in the way that plant life behaves and responds differently from animal life. For example, plant cells have a cell wall which restricts the shape of plant cells and thereby limits the flexibility and mobility of plants. Plant cells also contain chlorophyll, a chemical compound that interacts with light in a way that enables plants to manufacture their own nutrients rather than consuming other living things as animals do. Plant physiology deals with interactions between cells, tissues, and organs within a plant. Different cells and tissues are physically and chemically specialized to perform different functions. Roots and rhizoids function to anchor the plant and acquire minerals in the soil. Leaves catch light in order to manufacture nutrients. For both of these organs to remain living, minerals that the roots acquire must be transported to the leaves, and the nutrients manufactured in the leaves must be transported to the roots. Plants have developed a number of ways to achieve this transport, such as vascular tissue, and the functioning of the various modes of transport is studied by plant physiologists. Plant physiologists study the ways that plants control or regulate internal functions. Like animals, plants produce chemicals called hormones which are produced in one part of the plant to signal cells in another part of the plant to respond. Many flowering plants bloom at the appropriate time because of light-sensitive compounds that respond to the length of the night, a phenomenon known as photoperiodism. The ripening of fruit and loss of leaves in the winter are controlled in part by the production of the gas ethylene by the plant. Finally, plant physiology includes the study of how plants respond to conditions and variation in the environment, a field known as environmental physiology. Stress from water loss, changes in air chemistry, or crowding by other plants can lead to changes in the way a plant functions. These changes may be affected by genetic, chemical, and physical factors. Paradoxically, the subdiscipline of environmental physiology is on the one hand a recent field of study in plant ecology and on the other hand one of the oldest. Environmental physiology is the preferred name of the subdiscipline among plant physiologists, but it goes by a number of other names in the applied sciences. It is roughly synonymous with ecophysiology, crop ecology, horticulture and agronomy. The particular name applied to the subdiscipline is specific to the viewpoint and goals of research. Whatever name is applied, it deals with the ways in which plants respond to their environment and so overlaps with the field of ecology. Environmental physiologists examine plant response to physical factors such as radiation, temperature, fire, and wind. Of particular importance are water relations and the stress of drought or inundation, exchange of gases with the atmosphere, as well as the cycling of nutrients such as nitrogen and carbon.

This book explores the recent developments in plant physiology with regard to the demands of students and teachers in the field. Topics like metabolism, Photosynthesis, respiration, nutrition, water balance, plant growth, movements, resistance against unfavourable abiotic and biotic effects etc., are discussed. This book is highly recommended for both graduate and undergraduate students.

—Editor

Chapter

1

Plant-water Relations

Water is the most abundant constituent of all physiologically active plant cells. Leaves, for example, have water contents which lie mostly within a range of 55–85% of their fresh weight. Other relatively succulent parts of plants contain approximately the same proportion of water, and even such largely nonliving tissues as wood may be 30–60% water on a fresh-weight basis. The smallest water contents in living parts of plants occur mostly in dormant structures, such as mature seeds and spores. The great bulk of the water in any plant constitutes a unit system. This water is not in a static condition. Rather it is part of a hydrodynamic system, which in terrestrial plants involves absorption of water from the soil, its translocation throughout the plant, and its loss to the environment, principally in the process known as transpiration.

Cellular Water Relations

The typical mature, vacuolate plant cell constitutes a tiny osmotic system, and this idea is central to any concept of cellular water dynamics. Although the cell walls of most living plant cells are quite freely permeable to water and solutes, the cytoplasmic layer that lines the cell wall is more permeable to some substances than to others. If a plant cell in a flaccid condition—one in which the cell sap exerts no pressure against the encompassing cytoplasm and cell wall—is immersed in pure water, inward osmosis of water into the cell sap ensues. This gain of water results in the exertion of a turgor pressure against the protoplasm, which in turn is transmitted to the cell wall. This pressure also prevails throughout the mass of solution within the cell. If the cell wall is elastic, some expansion in the volume of the cell occurs as a result of this pressure, although in many kinds of cells this is relatively small.

If a turgid or partially turgid plant cell is immersed in a solution with a greater osmotic pressure than the cell sap, a gradual shrinkage in the volume of the cell ensues; the amount of shrinkage depends upon the kind of cell and its initial degree of turgidity. When the lower limit of cell wall elasticity is reached and there is continued loss of water from the cell sap, the protoplasmic layer begins to recede from the inner surface of the cell wall. Retreat of the protoplasm from the cell wall often continues until it has shrunk toward the centre of the cell, the space between the protoplasm and the cell wall becoming occupied by the bathing solution. This phenomenon is called plasmolysis.

In some kinds of plant cells movement of water occurs principally by the process of imbibition rather than osmosis. The swelling of dry seeds when immersed in water is a familiar example of this process.

Stomatal Mechanism

Various gases diffuse into and out of physiologically active plants. Those gases of greatest physiological significance are carbon dioxide, oxygen, and water vapour. The great bulk of the gaseous exchanges between a plant and its environment occurs through tiny pores in the epidermis that are called stomates. Although stomates occur on many aerial parts of plants, they are most characteristic of, and occur in greatest abundance in, leaves.

Transpiration Process

The term transpiration is used to designate the process whereby water vapour is lost from plants. Although basically an evaporation process, transpiration is complicated by other physical and physiological conditions prevailing in the plant. Whereas loss of water vapour can occur from any part of the plant which is exposed to the atmosphere, the great bulk of all transpiration occurs from the leaves. There are two kinds of foliar transpiration: (1) stomatal transpiration, in which water vapour loss occurs through the stomates, and (2) cuticular transpiration, which occurs directly from the outside surface of epidermal walls through the cuticle. In most species 90% or more of all foliar transpiration is of the stomatal type.

Transpiration is a necessary consequence of the relation of water to the anatomy of the plant, and especially to the anatomy of the leaves. Terrestrial green plants are dependent upon atmospheric carbon dioxide for their survival. In terrestrial vascular plants the principal carbon dioxide–absorbing surfaces are the moist mesophyll cells walls which bound the intercellular spaces in leaves. Ingress of carbon dioxide into these spaces occurs mostly by diffusion through open stomates. When the stomates are open, outward diffusion of water vapour unavoidably occurs, and such stomatal transpiration accounts for most of the water vapour loss from plants. Although transpiration is thus, in effect, an incidental phenomenon, it frequently has marked indirect effects on other physiological processes which occur in the plant because of its effects on the internal water relations of the plant.

Water Translocation

In terrestrial rooted plants practically all of the water which enters a plant is absorbed from the soil by the roots. The water thus absorbed is translocated to all parts of the plant. The mechanism of the "ascent of sap" (all translocated water contains at least traces of solutes) in plants, especially tall trees, was one of the first processes to excite the interest of plant physiologists.

The upward movement of water in plants occurs in the xylem, which, in the larger roots, trunks, and branches of trees and shrubs, is identical with the wood. In the trunks or larger branches of most kinds of trees, however, sap movement is restricted to a few of the outermost annual layers of wood.

Root pressure is generally considered to be one of the mechanisms of upward transport of water in plants. While it is undoubtedly true that root pressure does account for some upward movement of water in certain species of plants at some seasons, various considerations indicate that it can be only a secondary mechanism of water transport.

Upward translocation of water (actually a very dilute sap) is engendered by an increase in the negativity of water potential in the cells of apical organs of plants. Such increases in the negativity of water potentials occur most commonly in the mesophyll cells of leaves as a result of transpiration.

Water Absorption

The successively smaller branches of the root system of any plant terminate ultimately in the root tips, of which there may be thousands and often millions on a single plant. Most absorption of water occurs in the root tip regions, and especially in the root hair zone. Older portions of most roots become covered with cutinized or suberized layers through which only very limited quantities of water can pass.

Whenever the water potential in the peripheral root cells is less than that of the soil water, movement of water from the soil into the root cells occurs. There is some evidence that, under conditions of marked internal water stress, the tension generated in the xylem ducts will be propagated across the root to the peripheral cells. If this occurs, water potentials of greater negativity could develop in peripheral root cells than would otherwise be possible. The absorption mechanism would operate in fundamentally the same way whether or not the water in the root cells passed into a state of tension. The process just described, often called passive absorption, accounts for most of the absorption of water by terrestrial plants.

The phenomenon of root pressure represents another mechanism of the absorption of water. This mechanism is localized in the roots and is often called active absorption. Water absorption of this type only occurs when the rate of transpiration is low and the soil is relatively moist. Although the xylem sap is a relatively dilute solution, its osmotic pressure is usually great enough to engender a more negative water potential than usually exists in the soil water when the soil is relatively moist. A gradient of water potentials can thus be established, increasing in negativity across the epidermis, cortex, and other root tissues, along which the water can move laterally from the soil to the xylem.

Plant Mineral Nutrition

The relationship between plants and all chemical elements other than carbon, hydrogen, and oxygen in the environment. Plants obtain most of their mineral nutrients by extracting them from solution in the soil or the aquatic environment. Mineral nutrients are so called because most have been derived from the weathering of minerals of the Earth's crust. Nitrogen is exceptional in that little occurs in minerals: the primary source is gaseous nitrogen of the atmosphere.

Some of the mineral nutrients are essential for plant growth; others are toxic, and some absorbed by plants may play no role in metabolism. Many are also essential or toxic for the health and growth of animals using plants as food. Six basic facts have been established: (1) plants do not need any of the solid materials in the soil—they cannot even take them up; (2) plants do not need soil microorganisms;

(3) plant roots must have a supply of oxygen; (4) all plants require at least 14 mineral nutrients; (5) all of the essential mineral nutrients may be supplied to plants as simple ions of inorganic salts in solution; and (6) all of the essential nutrients must be supplied in adequate but nontoxic quantities. These facts provide a conceptually simple definition of and test for an essential mineral nutrient. A mineral nutrient is regarded as essential if, in its absence, a plant cannot complete its life cycle.

Nutrients which plants require in relatively large amounts, that is, the essential macronutrients, are nitrogen, sulfur, phosphorus, calcium, potassium, and magnesium. Iron is not required in large amounts and hence is regarded as an essential micronutrient or trace element. With the progressive development of better techniques for purifying water and salts, the list of essential nutrients for all plants has expanded to include boron, manganese, zinc, copper, molybdenum, and chlorine. Evidence has accumulated in support of nickel being essential. In addition, sodium and silicon have been shown to be essential for some plants, beneficial to some, and possibly of no benefit to others. Cobalt has also been shown to be essential for the growth of legumes when relying upon atmospheric nitrogen. Claims that two other chemical elements (vanadium and selenium) may be essential micronutrients have still to be firmly established.

Mineral nutrients may be toxic to plants either because the specific nutrient interferes with plant metabolism or because its concentration in combination with others in solution is excessive and interferes with the plant's water relations. Other chemical elements in the environment may also be toxic. High concentrations of salts in soil solutions or aquatic environments may depress their water potential to such an extent that plants cannot obtain sufficient water to germinate or grow. Some desert plants growing in saline soils can accumulate salt concentrations of 20–50% dry weight in their leaves without damage, but salt concentrations of only 1–2% can damage the leaves of many species.

A number of elements interfere directly with other aspects of plant metabolism. Sodium is thought to become toxic when it reaches concentrations in the cytoplasm that depress enzyme activity or damage the structure of organelles, while the toxicity of selenium is probably due to its interference in metabolism of amino acids and proteins. The ions of the heavy metals, cobalt, nickel, chromium, manganese, copper, and zinc are particularly toxic in low concentrations, especially when the concentration of calcium in solution is low; increasing calcium increases the plant's tolerance. Aluminium is toxic only in acid soils. Boron may be toxic in soils over a wide pH range, and is a serious problem for sensitive crops in regions where irrigation waters contain excessive boron or where the soils contain unusually high levels of boron.

All plants grow poorly on very acid soils (pH d" 3.5); some plants may grow reasonably well on somewhat less acid soils. Several factors may be involved, and their interactions with plant species are complex. The harmful effects of soil acidity in some areas have been exacerbated by industrial emissions resulting in acid rain and in deposition of substances which increase the acidity on further reaction in the soil, with consequent damage to plants and animals in these ecosystems.

The elemental composition of plants is important to the health and productivity of animals which graze them. With the exception of boron, all elements which are essential for plant growth are also essential for herbivorous mammals. Animals also require sodium, iodine, and selenium and, in the case of ruminant herbivores, cobalt. As a result, animals may suffer deficiencies of any one of this latter group of elements when ingesting plants which are quite healthy but contain low concentrations of these elements. In addition, nutrients in forage may be rendered unavailable to animals through a variety of factors that prevent their absorption from the gut. Plants and animals differ also in their tolerance of high levels of nutrients, sometimes with deleterious results for grazing animals. For example, the toxicity of high concentrations of selenium in plants to animals grazing them, known as selenosis, was recognized when the puzzling and long-known "alkali disease" and "blind staggers" in grazing livestock in parts of the Great Plains of North America were shown to be symptoms of chronic and acute selenium toxicity.

Essential Points in the Nutrition of Plants

The Nature of Plant Growth Processes

The processes concerned in the growth of plants are the subjects of study by plant physiologists and plant biochemists. A comprehensive account of these processes is outside the scope of the present work, the special object of which is to deal with the outward and visible signs of imperfections in the plant's activities caused by faulty mineral nutrition. Nevertheless it is useful to have before us the general features of the main processes involved and to realize that the symptoms we shall be discussing later have a physiological basis, and are not direct and unchangeable signs of the specific deficiencies but result from the derangement of the complicated mechanism of the plant's vital activities.

The main processes involved in plant development may be summarized as follows:

Absorption: Intake of water and mineral elements by the root system.

Carbon assimilation or photosynthesis: Intake of carbon dioxide from the air by the leaves, and reaction of the gas with water in the leaf in the presence of the green chlorophyll to form sugar and free oxygen.

Formation of protoplasm: Protoplasm is the living material of the plant, consisting mainly of proteins, complex compounds of nitrogen built up by the plant from more simple compounds of this element.

Respiration: The combination of oxygen with various food substances synthesized by the plant, especially sugars, whereby energy is produced.

Transpiration: Loss of water from the plant, mainly from the leaves.

Translocation: The movement of materials within the plant.

Storage: Storage of reserve products in various organs and tissues.

During growth there are continuous processes of building up of complex compounds of carbon and nitrogen and breaking down of these into more simple substances, in which water and oxygen are intimately concerned. These processes together comprise plant metabolism.

In the course of the metabolic processes innumerable substances are formed, such as sugars, starch, cellulose, acids, lignin, tannins, amino acids, proteins, amides etc., and many plants also produce special products, as for instance nicotine in the tobacco plant.

For the normal functioning of the above processes there must be an adequate intake of water by the plant to maintain the plant cells in a more or less turgid condition and, since water is being continuously lost at a varying rate from the plant, intake and movement within the plant tissues must bc capable of ready adjustment to these changes.

As a result of metabolic activities plants develop special organs of growth and reproduction, each of which has its special characters and makes particular demands on the nutrient supplies of the plant.

With all plants there are well defined seasonal growth cycles. Thus annuals, such as cereals, begin from the seed, give rise to seedlings, which later flower, form grain and ripen off, whilst perennial deciduous trees, such as apples, pears, etc., begin growth in the spring, using stored reserves of food, form leaves, make shoots, blossom and form fruits and subsequently shed their leaves, but meanwhile pass on reserve foods to various storage organs in preparation for the next season's growth. Coincident with these growth cycles there are well defined chemical cycles of nutrient elements and elaborated products in the leaves, stems and roots, etc. It will be shown later that these cycles are of great importance in considering p-deficiency effects and in diagnosing their causes.

The Plant Environment

Nutritional problems must be considered in relation to all the conditions in which plants live, and not merely in terms of the amounts of plant nutrients contained in or added to the soil. For example, those who are accustomed to growing plants know that the temperature must not be too low or no growth may result or that if too high the plants may be injured. An optimum temperature is usually recognized and this may vary according as to whether the plant is young or old. Similarly the importance of light is well known and plants may be put in special positions to obtain a maximum supply of light at one stage and may be shaded at another.

The actual duration of the daily period of illumination also affect growth and there are plants which are classified as requiring "long day" conditions to complete their growth cycles and others as needing "short day" conditions. If the special "long" or "short" day periods are not forthcoming for the respective classes of plants requiring these, their growth cycles are abnormal and they may fail entirely to produce flowers, grain or fruit.

The humidity of the atmosphere, as distinct from the water supply in the soil, is of importance in determining the water conditions within the plant, as these are dependent on both water intake by the roots and water loss from the leaves, the latter being largely influenced by the air humidity.

Even the presence of adequate quantities of plant nutrients in the soil is no guarantee that they will be absorbed by the plant roots. It will be shown later how these may be present in forms which are not available to the plants, but even when they would be considered as being present in suitable forms for absorption, other factors may prevent this taking place. An example of this latter condition is afforded in poorly aerated soils where lack of oxygen near the roots may prevent them from actively absorbing mineral nutrients

The problems of such influences in the plant environment as those just mentioned are complicated by the fact that they do not act independently, but their effects are modified by one another. Thus the effects of light intensity or period of daylight may vary with different temperature conditions.

The requirements of plants for different nutrients may be affected by conditions of light, temperature and water supply, and by other factors of the general environment. Thus the need for nitrogen may be less under conditions of relatively low light intensity whereas the need for potash in these circumstances may be greater, these facts being of importance in growing tomatoes under glass. The effect of nitrogen in relation to light may be shown by growing a plant under normal light conditions with insufficient nitrogen, when the leaves will show the well known symptoms of nitrogen deficiency-pale green, yellow, orange and red tints. If such a plant be then shaded, the leaves will turn a darker green and growth may be visibly increased. It can be shown that the lowered light conditions result in an increase of "soluble" a breaking down of proteins, thereby rendering the nitrogen of these available for growth processes.

This interrelationship of environmental factors is well illustrated by an experiment on apple trees at Long Ashton.

Bramley's Seedling trees were grown in compost in large pots and given a small dressing of a nitrogenous fertilizer. Some of the trees were grown in a specially constructed glass house and an equal number in an adjoining wire enclosure. The trees in the enclosure showed severe symptoms of nitrogen deficiency-pale green and yellow leaves, reddish brown barks and highly coloured, red fruits. The condition was corrected by further dressings of nitrogen. In contrast, the trees under glass, where the light was of less intensity and the temperature higher, made vigorous growth, carried large, green leaves and bore large, green fruits.

Iron and zinc deficiency symptoms may be less severe under conditions of low light intensity, whilst boron deficiency effects are less severe and magnesium deficiency effects are more pronounced in wet seasons than in dry ones.

The rate of water absorption is less at lower temperatures than at higher ones and efficient intake is also dependent on good aeration. These facts may result in a water deficit within plants growing in cold, wet soils when the air temperature is high. The raw materials needed for plant growth consist of carbon dioxide, which is obtained from the atmosphere through the stomata of the leaves, and water and the so-called mineral nutrients, which normally enter the plant through the medium of the roots.

The importance of water and carbon dioxide in the nutrition of plants will be apparent from the facts that water often comprises 80 to 90 % of the total weight of growing plants, and carbon and oxygen together may account for over 80% of their dry matter, i.e., the solid matter remaining after water is removed. As against these large amounts, the mineral nutrients, as measured by the ash content of the plants, i.e., the mineral residue obtained when the organic matter is destroyed by heat, often contribute from 5 to 15% of the dry matter.

It has been shown in recent years that certain organic compounds, known as "growth promoting substances" or "hormones", which occur in plants, and some of which are also present in soils and natural manures, are capable of producing marked growth responses, such as increased root growth, shoot and leaf curvatures, stimulation or suppression of buds, increased fruit setting, prevention of fruit abscission etc.

They appear to perform important functions in the growth of plants. Examples of substances of this kind which can produce growth responses are 13 indole-acetic acid, 13 indole-butyric acid, phenyl acetic acid, A, naphthalene-acetamide, vitamin B_1

It is not at present clear to what extent growth substances are absorbed by plants from soils, although it has been shown that vitamin B_1, which occurs naturally in soils, can be obtained in this way.

The Mineral Nutrients

It has been shown in numerous researches that certain elements are necessary for the healthy growth of plants. They are sometimes spoken of as essential elements and, since some are needed in relatively large quantities and others in very small amounts, the former are referred to as "major" elements and the latter as "minor" or "trace" elements, or as micro nutrients.

The terms "major" and "minor" do not refer to the relative importance of the functions of the elements in plant growth, and for this reason the term "trace" element is preferable for the latter class.

Major elements: Nitrogen, phosphorus, calcium, magnesium, potassium, sulfur.

Trace elements: Iron, manganese, boron, copper, zinc and molybdenum.

(Iron occupies an intermediate position and is usually included in the major elements group. In dealing with field problems it is more convenient to group it with the trace elements.)

In addition, there are, other elements, such as sodium, chlorine and silicon, which produce beneficial effects on the growth of certain plants but which have not so far been shown to be absolutely essential to growth. The element aluminium is of general occurrence in plants, but seems to be without direct nutritional value, although aluminium sulfate is used, because of its acidifying properties, to change the colour of hydrangeas growing on alkaline soils from pink to blue, and aluminium may also exert indirect influences on nutritional processes.

Other elements often occur in plants but they are not known to serve any useful function and frequently they act as plant poisons or toxins.

The nutrient elements can only be absorbed by plants when present in certain forms: nitrogen from nitrates and ammonium salts; phosphorus from phosphates; calcium, magnesium and potassium from their salts; sulfur from sulfates; iron from ferrous or ferric salts (more readily from ferrous salts); manganese from manganous salts; boron from borates; copper and zinc from their salts, and molybdenum from molybdates.

There may appear to be certain exceptions to this statement in practice. For instance, nitrogen may be applied to a soil as "organic" nitrogen, as in hoof meal or urea, and sulfur may be added as the element itself, as in flowers of sulfur, ground sulfur, etc. In such conditions the added materials are, however, converted into the nitrate and sulfate forms respectively by soil organisms before being absorbed by the plants.

Further points of importance in connection with the absorption of the mineral nutrients by plants are as follows:

(a) They must be absorbed from relatively dilute solutions or the plants will be! injured or even killed.

(b) Certain of the elements slow down the absorption of others into the plant, e.g., calcium slows down potassium and vice versa. The phenomenon is known as "antagonism".

(c) Healthy plants result when the nutrients are absorbed in certain relative proportions. When the proportions are suitable the nutrient medium is said to be "balanced". When ratios between nutrients are too wide, deficiency conditions are created. Thus if a high proportion of nitrogen to potassium is absorbed, the plant will suffer from potassium deficiency.

(d) Nutrients, even though present in the nutrient solution in satisfactory amounts and proportions, may not be absorbed by the plant unless the "reaction" of the solution as regards acidity and alkalinity is satisfactory. The reaction is measured in terms of the pH scale, which is merely a convenient notation for stating the conditions of acidity in the solution (strength or intensity of acidity, not total amount). The neutral point is represented by pH 7.0; below this value the solution is acid and above it is alkaline. Many crop plants prefer a reaction slightly on the acid side-pH 6.0 to 6.5 and extreme values are in the neighborhood of 4.0 on the acid side and 9.0 on the alkaline side.

(e) The nutrient medium must contain an adequate supply of oxygen, i.e., aeration must be satisfactory.

The Functions of The Mineral Nutrients

Knowledge of the main functions of the mineral nutrients is useful in helping us to understand the effects produced by deficiencies of any one of them.

Essential Major Elements

Nitrogen. Nitrogen is a major constituent of several of the most important substances, which occur in plants. It is of outstanding importance among the essential elements in that nitrogen compounds comprise from 40 to 50% of the dry matter of protoplasm, the living substance of plant cells. For this reason nitrogen is required in relatively large quantities in connection with all growth processes in plants. It follows directly from this that without an adequate supply of nitrogen appreciable growth cannot take place and that plants must remain stunted and relatively undeveloped when nitrogen is deficient.

Proteins, which are of great importance in many plant organs, e.g., seeds, are compounds of nitrogen whilst chlorophyll, the green colouring matter of the leaves, also contains the element.

From this latter fact it will be apparent that when nitrogen is deficient leaves will contain relatively little chlorophyll, and will thus tend to be pale green in colour. In addition to the above substances, numerous other organic compounds of importance in plants, such as amino acids, amides, and alkaloids, are compounds of nitrogen.

Certain compounds of nitrogen are very mobile in plants, and this enables them readily to mobilize supplies of the element at vital growing points and to transfer stored supplies to points where they are most required. Such transference is common from old tissues to young growing points when supplies of the element are short. This mobility and re-utilization of nitrogen explains why deficiency symptoms of the element always appear first in the older parts of plants and why growing points are the last to be affected.

Phosphorus: This element, like nitrogen, is closely concerned with the vital growth processes in plants as it is a constituent of nucleic acid, and nuclei in which this occurs are essential parts of all living cells. Hence a deficiency of this element will also be expected to result in greatly restricted growth. Phosphorus is also of importance in seeds and in connection with the metabolism of fats. Compounds of phosphorus are concerned with the processes of respiration and with the efficient functioning and utilization of nitrogen.

This relationship to nitrogen probably accounts for the fact that several of the symptoms of phosphorus deficiency are identical or similar to those which result from a deficiency of nitrogen, Phosphorus is also of special importance in the processes concerned in root development and the ripening of seeds and fruits.

Calcium: Calcium occurs in plants chiefly in the leaves and the amounts present in seeds and fruits are relatively low. One of its main functions is as a constituent of the cell wall, the middle lamella of which consists largely of calcium pectate. This function appears to be of fundamental importance since, if calcium is replaced by any other of the essential elements, such as magnesium or potassium, the organic materials and mineral salts in the cells are readily leached through the walls.

Other functions attributed to calcium are as follows:

It provides a base for the neutralization of organic acids; it is concerned with activities of growing points (meristems), especially with root tips; it may be of importance in nitrogen absorption.

Although a large proportion of the calcium contained in the plant may be soluble in water-as much as 60% in cabbage calcium does not appear to move freely from the older to the younger parts of plants, and hence young tissues contain lower proportions of calcium than older ones. This may explain why calcium deficiency effects begin at the tips of shoots.

Magnesium: The outstanding fact about magnesium is that it is a constituent of chlorophyll, and is essential to the formation of this pigment. As a result, when magnesium is deficient, one of the symptoms commonly shown by plants is chlorosis. Magnesium is also regarded as a carrier of phosphorus in the plant, particularly in connection with the formation of seeds of high oil content, which contain the compound lecithin.

The element seems to be very mobile within the plant, and when deficient is apparently transferred from older to younger tissues where it can be re-utilized in the growth processes. This agrees with the observation that signs of magnesium deficiency invariably make their appearance first on the oldest leaves and progress systematically from them towards the youngest ones.

Potassium: Unlike all the other major elements, potassium does not enter into the composition of any of the important plant constituents, such as proteins, chlorophyll, fats and carbohydrates, concerned in plant metabolism. For this reason its role is more difficult to determine, and in spite of much study it cannot be said that the functions of potassium are clearly understood.

The element is present in all parts of plants in large or fairly large proportions. It seems to be of special importance in leaves and at growing points, as these are especially rich in potassium. Probably the whole of the potassium in plants is present in soluble form, and most of it seems to be contained in the cell sap and cytoplasm.

It is outstanding among the nutrient elements for its mobility and solubility within the plant tissues, and these properties no doubt account for the ready way in which potassium can be re-utilized by young tissues when the element is in short supply.

Among the functions which have been attributed to potassium and the processes with which it may be concerned, the following may be mentioned: The formation of carbohydrates and proteins; the regulation of water conditions within the plant cell and of water loss by transpiration; as a catalyst and condensing agent of complex substances; as an accelerator of enzyme action; as contributing to photosynthesis through its radioactive properties.

It has been shown in many instances that the potassium content of plants is frequently much higher than is necessary for healthy growth, and it is generally considered that luxury absorption of potassium often takes place.

The great mobility of potassium in plants, its special importance for and its reutilization by young tissues, and its apparent functions as a regulator of plant processes on a large scale are in harmony with the observations that, when potassium is moderately deficient, the effects are seen first in the older tissues and progress from these towards the growing points, but, when the deficiency is acute, growing points are severely affected, and die-back and general collapse of the plants commonly occur.

Sulfur: Sulfur occurs in plants as a constituent of proteins, and of certain volatile compounds such as mustard oil. It seems to be connected with chlorophyll formation, although it is not a constituent of this substance. Its functions in connection with proteins and chlorophyll doubtless account for the similarity of its deficiency effects to those due to deficiency of nitrogen.

Essential Trace Elements

Iron: Iron is closely concerned with chlorophyll formation but is not a constituent of it. Its role appears in this connection to be that of a catalyst. As a result of this function of iron, chlorosis is invariably an outstanding symptom when the element is deficient. Iron may also act as a catalyst, in the role of an oxygen carrier, in respiration.

A point of great importance in connection with iron is its relative immobility in plant tissues. Its mobility seems to be affected by several factors, such as the presence of manganese, potassium deficiency and high light intensity. There is evidence that the amount of chlorophyll is related to "active" iron in plants.

It will thus be seen that so-called iron deficiency in the plant may in fact usually mean iron immobility. Lack of mobility may also account for the fact that iron deficiency is first shown in the younger tissues.

Manganese: The functions of manganese are regarded as being closely associated with those of iron and as being concerned with chlorophyll formation. Hence, when manganese is deficient, chlorosis is a common symptom. Manganese may decrease the solubility of iron by oxidation and hence, an abundance of manganese within the plant may lead to iron deficiency and chlorosis.

Manganese is regarded as having the functions of a catalyst; its activities being specially concerned with oxidation and reduction reactions within the plant tissues.

Boron: The exact role of boron is not known, but again the evidence as for the other trace elements, suggests that, its functions are those of a catalyst or reaction regulator. Particular effects attributed to boron are as follows:

It can delay the onset of calcium deficiency effects but cannot replace calcium; it tends to keep calcium soluble; it may act as a regulator of potassium/calcium ratios, and of the absorption of nitrogen; it may be concerned with the oxidation-reduction equilibrium in cells.

Such functions as the above accord with the results which follow from a deficiency of the element, when growth processes show sudden collapse and drastic derangements of metabolism occur.

Zinc and Copper: Although specific functions have not been determined for these elements, here again the evidence points to their roles as catalysts and regulators. Deficiencies of both are associated with chlorosis and a serious general collapse of vital growth processes.

Since catalysts are not used up in the chemical reactions which they promote, we can understand how it comes about that quite small or even minute quantities of the "trace elements", iron, manganese, boron, zinc and copper, may nevertheless be essential to the plant's health and growth.

Elements Sometimes Beneficial

Sodium: As sodium is not strictly an essential element it cannot be expected to have a specific role in the metabolic activities of plants.

Where sodium produces significant effects it is often regarded as a conserver of potassium and as being able partly to replace that element in its role. In no instance, however, has it been shown that sodium can wholly replace potassium where the latter is acutely deficient.

In such circumstances, sodium is ineffective as a substitute for potassium, even for sodium-loving plants, such as sugar beet, mangold and barley. Sodium seems to affect the water relations of plants and often enables sugar beet and other crops to withstand drought conditions which would otherwise produce severe adverse effects.

Chlorine: The evidence of the role of chlorine in plants is somewhat contradictory, and no general statement can be made. In tobacco it has been shown to increase the water content of the tissues and to affect carbohydrate metabolism, leading to an accumulation of starch in the leaves. The element is present in plants as chloride, and is wholly soluble.

Toxic Effects of Mineral Elements

Toxic effects in plants may be produced both by essential nutrient elements and non-essential elements.

In the first class, the major nutrients are much less toxic than the trace elements. Indeed, for the major nutrients, there exists a fair safety margin for excess or "luxury" consumption, but for the trace elements the margin is very narrow. Similar conditions exist in relation to non-essential elements thus some plants will tolerate fairly large amounts of elements such as sodium or chlorine, but are injured by relatively small amounts of elements like arsenic or chromium.

Two types of injury may occur: (1) An excess of one element may lead to a deficiency of another which ultimately results in a deranged metabolism, e.g., excess nitrogen or excess phosphorus may result in insufficient potassium and excess potassium may lead to deficiency of magnesium or calcium. This type of injury applies particularly to essential nutrients. (2) The presence of an element may directly injure the protoplasm and bring about the speedy death of the plant.

Signs of Faulty Nutrition in Plants

When plants are grown in unsuitable environments, including conditions of faulty mineral nutrition, they react to the particular defects in more or less specific ways. Thus, if light is insufficient, the green colouring matter of the leaf will be lacking and the leaves may be almost white (chlorotic), and the plant may be very spindly and "drawn" in appearance; if the temperature is too high, the growth may be rank and soft; if water is insufficient, growth may be restricted, the tissues woody and the green of the leaves show a bluish tint.

Again, deficiencies and excesses of the individual elements produce characteristic effects on various organs of plants: foliage characters, including colour, density, size and shape of leaves; stem characters, such as thickness, colour and length of inter-nodes; root characters, such as colour, amount of fibre, abnormal thickening; blossom characters, including amount and time of opening of the flowers; fruit characters, such as size, colour, hardness and flavour.

Ability to recognize these particular effects forms the basis of the visual method of diagnosing plant deficiencies. Many of them can be readily learned and applied by practical farmers. Indeed, for many years progressive fruit growers in this country have used the leaf symptoms of deficiencies of nitrogen and potassium, and more lately of magnesium, as the main guide to manuring their fruit trees and bushes with these elements.

Detailed descriptions of deficiency symptoms and of the methods of using them in the field for diagnosing the manurial needs of crops.

Soils in Relation to the Supply of Mineral Nutrients

Since plants obtain their supplies of mineral nutrients from the soil, it is necessary to have some knowledge of soil conditions in order to understand the main problems of the supply of nutrient elements.

The Constitution of Soils

One of the first points which is brought out in any examination of a number of soils is their great diversity. In practical terms, soils may be good or bad; light or heavy; wet or dry; fertile or hungry; deep or shallow; black, red or brown; peaty, sandy, silty or clay; all terms denoting points of importance in relation to fundamental characters and to their practical utilization.

Soils are very complex bodies but, although showing such great variation, they all consist of five main components, mineral matter, organic matter, soil water, the soil atmosphere and a population of microorganisms.

Mineral Matter: The mineral matter furnishes the skeleton of the soil. It consists of materials which range in size from rock fragments and large pebbles to minute particles of clay which can be suspended in water for considerable periods. The mineral portion, which largely determines the texture of the soil, can be separated into its component parts or "fractions" by a combination of sieving and sedimentation in water. The British standard method of grading is as follows:

Fraction	*Diameter Limits*
Stones	> 2.0 m.m.
Coarse Sand	2.0—0.2 m.m.
Fine Sand	0.2—0.02 m.m.
Silt	0.02—0.0002 m.m.
Clay	< 0.002 m.m.

A preponderance of the coarser fractions tends to make soils light, free working, freely drained and hungry, whilst where the finer fractions predominate (especially clay) the soils are heavy, sticky, difficult to work and retentive of water and manures.

The clay fraction is of special importance as it possesses colloidal properties, which give to soils many of their characteristic properties, such as swelling and shrinkage, holding up of water and absorption of mineral nutrients from manures. The clay colloid, together with the organic colloid (humus), acts as the soil storehouse for available plant nutrients.

The mineral matter of soils provides the main natural source of mineral nutrients, which become available to plants through the weathering of rock minerals and the complex processes of soil formation.

Organic Matter: The content of organic matter in soils shows great variation. Organic matter may comprise almost the whole of the solid matter of peats, while in mineral soils it may account for only 1 or 2% of the soil. It is usual in mineral soils in this country for the organic matter to range from 3 to 10% in the soil, the lower values occurring in arable soils and the higher ones in old gardens and under grass covers in pastures and meadows.

Plant residues, either naturally or artificially added, provide the source of organic matter in soils. These residues vary greatly in character and composition, depending on the plants from which they are derived, whether from trees, grasses, clovers, cereals, root and vegetable crops, etc., and on the nutrient supplying powers of the soils on which they are grown. In addition to plant residues, soil organisms and animal remains contribute to the organic matter. In the soil the fresh material undergoes chemical change, especially by the action of soil organisms, and when this has proceeded to the stage where the original cellular structure is no longer recognizable we speak of the brownish product as humus. The organic matter in the soil at any time thus consists of fresh and partly decomposed residues and humus.

The main points in the formation of humus are as follows:

The raw residues contain a variety of compounds, such as proteins and other nitrogenous materials, and sugars, starch, cellulose, tannins, lignin and nutrient salts. The soil organisms use compounds like sugars (carbohydrates) as sources of energy, ultimately oxidizing them to carbon dioxide. To do this they require a supply of nitrogen which they way take either from the nitrogenous compounds in the residues, or if this supply is insufficient they may use other nitrogen present in the soil. The most resistant to their attack of the compounds contained in the residues appears to be lignin, and it seems probable that the end-point of the more rapid decomposition processes humus is a complex mixture in which lignin derivatives preponderate.

During these processes the ratio of carbon to nitrogen (C/N) in the residues alters from about 40/1 to 10/1 so that the process entails a large reduction in carbon.

It will thus be seen that the organic matter of the soil supports the population of soil organisms, and of this population the bacteria alone may be of the order of 20 to 40 millions per gram of soil.

These organisms play a very great part in determining the availability of mineral nutrients in soils, by breaking down plant residues and also by providing carbon dioxide which, in combination with water, is of great importance for the weathering of the soil minerals. Certain of the soil organisms, especially bacteria, are also able to fix free nitrogen from the air and thus enrich the soil with this element. Other bacteria are present which cause the soil to lose nitrogen, and there are multitudes of protozoa which feed on the bacteria, and if too abundant render the soil unhealthy. A properly balanced soil population is requisite for good crops and disturbances of this balance either in a beneficial or deleterious direction may be brought about by soil treatment.

It is important to realize, however, that bacteria and other soil organisms do not work for the special benefit of crops, and that the changes they bring about are to meet their own particular needs, which are not always in harmony with the requirements of the crops. This point can be illustrated by digging into soil fresh straw or sugar and growing a crop on it immediately, when the crop will most probably show symptoms of acute nitrogen deficiency due to the fact that, in order to consume the large amount of carbohydrate contained in the straw or sugar, the organisms deplete the soil of its readily available nitrogen.

Organic matter produces both chemical and physical effects on soils. Like clay, the humified material possesses colloidal properties, depending on the minute size of the constituent particles and the enormous surface they collectively present. It can swell and shrink and absorb nutrient salts. Together with clay, humus coats the mineral particles and can bind these together forming "crumbs", thus giving "structure" to soils. Humus acts as a binding agent for the particles of coarse sandy soils and exerts an "opening" effect on close-textured clay soils. The binding properties are not associated with stickiness to the same extent as those of clay. Humus is superior to clay colloid as a storehouse of nutrient bases.

Soil Water: The solid particles of the soil, consisting of the mineral particles and the organic matter, being of varying dimensions, have spaces of varying sizes between them which are occupied by the soil water and the soil atmosphere. The total volume of these spaces in any soil is known as the pore space. It varies for different soils and even in the same soil under influences such as rainfall, drought, frost and management, but usually occupies from 30 to 60% of the soil volume. It will be clear that the amounts of water and air in the pore space are complementary and that when the soil is wet there will be correspondingly little air present and vice versa. A high water content in the soil thus means a condition of relatively poor aeration.

The manner in which water is held in soils and the way in which it moves through the soil mass provide difficult scientific problems. There is general agreement that in a wet soil a certain proportion of the water can be readily removed, and that some is held very strongly by forces in the soil. Moreover, it can be shown that only a proportion of the water in soils is available to plants. The movement of water through soils is also limited, and water tables at considerable depths below the surface cannot be relied upon to supply water to the surface layers of the soil by the well known capillary action which occurs in narrow tubes. Similarly, if a dry soil is wetted from the surface, the whole of the soil to a considerable depth does not become wetted through uniformly, but the top layers only are wetted, practically to full capacity, and the layers below remain dry.

This strictly limited movement of water through soils is of great practical importance, since it means that for practical purposes water does not move appreciably to the plant roots but the roots must go after the water.

The soil water contains the soluble products of the soil, and is the main nutrient medium for the plant roots. As such it is commonly called the soil solution. There is evidence that plants may also feed directly on nutrients contained in the soil colloids.

The soil solution is very dilute. Measurements made on displaced solutions show concentrations of soluble materials between 0.1 and 1.0%, whilst if drainage waters are analysed the corresponding values are of the order of 0.02 to 0.5%. In both, the proportions of calcium, nitrate, sulfate, carbon dioxide, silica and organic matter are relatively high, and there may be large proportions of chloride, sodium and magnesium, but potassium is always low and phosphate and ammonia, are present only in traces.

The soil organisms play an important part in providing soluble products for the soil solution. They liberate into the solution nitrate, sulfate and carbon dioxide and these, acting as acids, dissolve corresponding or equivalent amounts of base-forming elements, particularly calcium. Further changes then occur, the calcium in particular bringing into solution other "bases" such as magnesium and potassium by "base" exchange reactions with the colloidal clay and humus materials.

Under normal conditions, the whole of the nitrate and chloride is present in the soil solution, and they pass into the drainage water if not quickly absorbed by plants. Sodium is also mainly present in soluble form. Calcium, magnesium and potassium are held in considerable amounts in the soil colloids as "exchangeable bases", and in this form are less easily washed out of the soil.

Phosphorus is mostly present as insoluble compounds, as calcium phosphates in neutral and alkaline soils, and as iron phosphate in acid soils. Both potassium and phosphorus are only present in the soil solution in very small amounts, but apparently supplies of these elements in the solution are quickly replenished when removed by plants.

The amounts of iron, aluminium and manganese dissolved are largely dependent on the soil reaction (pH) and in strongly acid soils they may assume toxic concentrations to plants.

The conditions governing boron solubility are not known though, under drought conditions and at high pH values, availability to plants is reduced.

Copper and zinc (perhaps also manganese) appear to present complicated problems in which organic matter and the soil organisms are concerned.

Molybdenum may be more readily soluble in soils under alkaline than acid conditions, as it has been shown to be more readily available to pasture plants under the former conditions.

The Soil Atmosphere: The soil atmosphere or soil air, near the surface, has a composition similar to that of the ordinary air and diffusion between the two is rapid. Except near the surface the soil air is saturated with moisture. The carbon dioxide content may fluctuate considerably, being increased by organic matter, by cropping, by high temperatures, which speed up the activities of soil organisms, and by the respiration of plant roots.

In poorly drained soils the concentration of carbon dioxide may be very high, with a correspondingly low oxygen content, and in extreme condition this may actually lead to inefficient absorption and even to root injury due to lack of oxygen. The fact that carbon dioxide accumulates and oxygen supply is depleted under conditions of poor drainage is important in considering deficiency problems in wet soils where practical results often appear contrary to expectation.

Availability of Mineral Nutrients in Soils

The mineral elements available to plants in soils may be present as the result of natural soil processes or may come from additions of natural manures and fertilizers.

Nitrogen: Nitrogen availability is intimately connected with the activities of soil organisms. Nitrate is the normal form in which nitrogen is absorbed by plants from soils, although ammonia can also be utilized. Nitrogen in organic compounds is converted into nitrate by a chain of reactions brought about by organisms. For protein nitrogen this may be represented as follows:

Protein → intermediate organic products, such as amino acids → ammonia → nitrite → nitrate.

The effect of excess carbohydrate on these reactions has been mentioned.

Ammonia may be held in the colloid complex as an "exchangeable base" in the same way as calcium, magnesium and potassium.

Phosphorus: This element exists in the soil in many forms, both as organic and inorganic compounds, and is also added to the soil in manures and fertilizers in a variety of materials, such as in meat meal, bone manures, basic slag, ground phosphate rock (mineral phosphate) and water-soluble superphosphates. Evidence indicates that water-soluble forms are generally most readily available to plants even though they are rendered insoluble almost immediately after application to the soil. Organic forms of phosphorus are usually less readily available than the inorganic compounds. Phosphates undergo many changes in soils both by organisms and by purely chemical reactions and, even though very heavy dressings are applied, the amount of water-soluble phosphate in the soil at any time is very small.

The movement of phosphates in soils is very limited, and soils are said to have high "fixing powers" for phosphates. Heavy soils, as a rule, show higher fixing powers than light ones, and soils with high iron contents possess specially strong fixing properties. The two elements mainly responsible for the fixation of phosphates are calcium, in neutral and alkaline soils, and iron in acid soils.

Table: *The Relative Fixing Powers of some Light and Heavy Soils for Phosphates*

	Percent Phosphoric Acid (P_2O_5) soluble in 1% Citric Acid Solution.							
	Light Sands				***Clays***			
Surface Soil	0.0600	0.0865	0.1472	0.1730	0.0485	0.0601	0.0810	0.1158
Subsoil	0.0230	0.0616	0.0932	0.0734	0.0011	0.0085	0.0038	0.0120

Important practical points in connection with the fixation of phosphates are that a large proportion of the phosphates added to many soils never become available to the crops, and that phosphates should always be placed as near as possible to the roots of the plants for which they are intended.

In some soils where fruit trees suffer from phosphate deficiency, the method of injecting the phosphatic materials into the soils through high-pressure lances has been suggested to obtain local concentrations of fertilizer near portions of the roots. For cereal crops the difficulty has been met by the introduction of combined seed and fertilizer drills, which deliver the seed and phosphate into the soil in close proximity, and by the use of granular fertilizers, of which only a small proportion aof the material comes into contact with the soil.

Calcium: Calcium occurs in soils in a large variety of minerals, and carbonate of lime (calcium carbonate) may comprise a large percentage of the mineral matter where the soil is derived from limestone or chalk rocks. The element is readily leached from the soil and, on sandy soils where it is not abundant, frequent replenishment by means of lime or limestone dressings is necessary.

Carbonate of lime is readily brought into solution in soils by means of carbon dioxide dissolved in the soil water and thus, if the soil contains lime, a supply of calcium in soluble form is readily assured. Calcium also comprises the major proportion of the elements held as exchangeable bases where soils are not strongly acid and it is readily brought into the soil solution from this state.

Magnesium: The mode of occurrence of magnesium in soils is fairly similar to that of calcium. It occurs as carbonate and in a variety of minerals. Like calcium it is readily brought into the soil solution from the carbonate, and is held in soils as an exchangeable base. It is fairly easily leached, and for this reason may become deficient in sandy soils during wet periods.

Potassium: Potassium is widely distributed in soil minerals, such as potash-felspar, mica and glauconite, from which it is slowly converted into soluble forms by weathering processes. Heavy soils contain higher amounts of potassium than light soils. Potassium is fairly strongly "fixed" in soils, possibly largely as an exchangeable base. Very small amounts of potassium are present in the soil solution at any given time, but exchangeable potassium appears to be readily available to plants.

Sodium: Sodium is brought into solution from minerals in a similar manner to potassium, but it remains mainly in the soil solution, being only weakly absorbed by the colloids, and is very readily leached from the soil. In salt marshes recently reclaimed from the sea and in alkali soils, where sodium is the principal base-forming element, it is the main exchangeable base. Under these conditions the soils are often unfit for cultivation, due to bad physical condition, until the sodium has been replaced by calcium.

Soluble sodium compounds are readily available to plants.

Sulfur: Sulfur is present in soils both in organic and inorganic forms. Inorganic sulfur is mainly present as sulfate, but sulfide may accumulate where conditions are favourable for "reducing" reactions to take place.

Under water-logging conditions, calcium sulfate may form crystalline deposits in the subsoil, and this substance may also be seen as a white deposit on the surface of soils when manuring is high, as in market gardens.

Sulfur compounds are changed from one form to another in the soil by special bacteria, the end product of whose reactions is the sulfate form where conditions favour oxidation. Thus, if the element sulfur is added as a dressing to soil, it may be quickly oxidized to sulfuric acid, which reaction is used in treating alkaline soils to lower the pH value.

Sulfate: Sulfur is an important constituent of the soil solution, and in this form sulfur moves readily through the soil.

The widespread occurrence of sulfur in rocks and organic matter and its mobility in the soil doubtless account for the rarity of a deficiency of the element in crops.

Chlorine: This element is of common occurrence in soils in the form of chloride, and as such moves freely through the soil in the soil solution, from which it is available to plants.

Common salt or sodium chloride is. also present in considerable quantity in the atmosphere near the sea, and a proportion of this reaches the soil, adding both chloride and sodium.

Iron: Iron is present in most soils, mainly in the form of its oxides, which are largely responsible for the red and brown colours in soils. Where drainage and aeration are good, ferric compounds predominate, and where water-logging and bad aeration occur ferrous compounds are formed. The insoluble forms of iron are brought into solution by the action of acids. Organic compounds of iron, such as the humate, appear to be mobile in the soil. The availability of iron to plants increases with acidity and is depressed by phosphates.

Manganese: The mode of occurrence of manganese in the soil resembles that of iron in that the oxides are important forms. Both iron and manganese are intimately concerned in the oxidizing and reducing reactions which take place in soils, and compounds of both in the form of concretions appear in imperfectly drained soils.

The more highly oxidized compounds of manganese, such as manganese dioxide, are of very low availability to plants. Like iron, the solubility of the soil manganese increases with increasing acidity, and in many soils it is largely unavailable to plants above pH 6.5.

The availability of manganese seems also to be greatly affected by organic matter and drainage conditions, deficiency of the element being common on calcareous peats and other soils with high organic matter content, and on soils with high water tables.

Boron: Little is known of the sources of boron available to plants in soils. The element is a constituent of the mineral tourmaline, which occurs in many rocks, but this is very resistant to weathering. When applied to the soil in the form of a borate, boron appears to move fairly readily through the soil. The availability of boron to the plant is decreased by liming and by dry soil conditions. In practice over-liming is a common cause of boron deficiency.

Zinc and Copper: In the neighborhood of zinc or copper mineral deposits the soils are generally toxic to crops. The contents of these elements in normal soils are very small, and maybe are largely the result of concentrations and additions from growing plants and added residues.

Little is known of the factors affecting their availability to plants, but organic matter and soil organisms may be important factors.

Chapter

2

Photosynthesis: The Light Reactions

Principles of Spectrophotometry

Much of what we know about the photosynthetic apparatus was learned through spectroscopy—that is, measurements of the interaction of light and molecules. Spectrophotometry is an important branch of spectroscopy that focuses on the technique of measurement. Here we will examine four topics: Beer's law, the measurement of absorbance, action spectra, and difference spectra.

Beer's Law

An essential piece of information about any molecular species is how much of it is present. Quantitative measures of concentration are one of the cornerstones of biological science. Of all the methods that have been devised for measuring concentration, by far the most widely applied is absorption spectrophotometry. In this technique, the amount of light that a sample absorbs at a particular wavelength is measured and used to determine the concentration of the sample by comparison with appropriate standards or reference data. The most useful measure of light absorption is the absorbance (A), also commonly called the optical density (OD). The absorbance is defined as $A = \log I_0 / I$ where I_0 is the intensity of light that is incident on the sample and I is the intensity of light that is transmitted by the sample.

The absorbance of a sample can be related to the concentration of the absorbing species through Beer's law:

$$A = \varepsilon cl$$

where c is concentration, usually measured in moles per litter; l is the length of the light path, usually 1 cm; and ε is a proportionality constant known as the molar extinction coefficient, with the units of litres per mole per centimetre. The value of ε is a function of both the particular compound being measured and the wavelength. Chlorophylls typically have an ε value of about 100,000 L mol^{-1} cm^{-1}. When more than one component of a complex mixture absorbs at a given wavelength, the absorbances due to the individual components are generally additive.

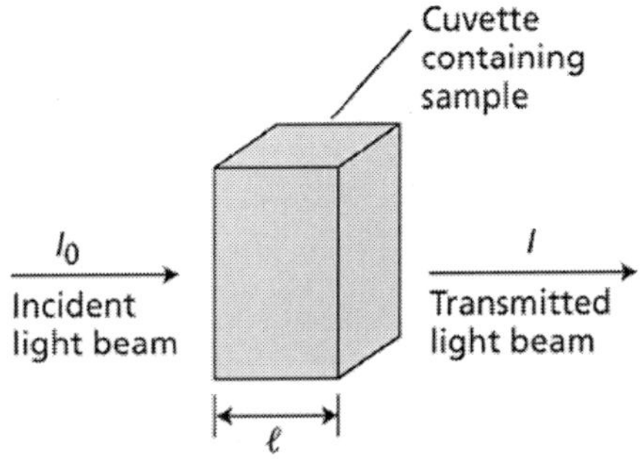

***Figure:** Definition of absorbance. A monochromatic incident light beam of intensity I_0 traverses a sample contained in a cuvette of length (l). Some of the light is absorbed by the chromophores in the sample, and the intensity of light that emerges is I.*

The Spectrophotometer

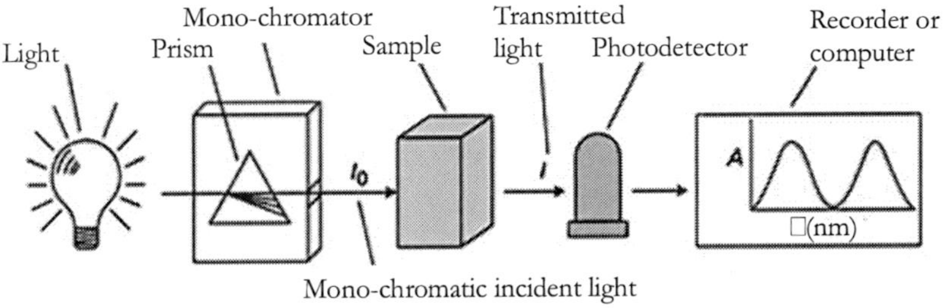

***Figure:** Schematic diagram of a spectrophotometer. The instrument consists of a light source, a monochromator that contains a wavelength selection device such as a prism, a sample holder, a photodetector, and a recorder or computer. The output wavelength of the monochromator can be changed by rotation of the prism; the graph of absorbance versus wavelength is called a spectrum.*

The absorbance is measured by an instrument called a spectrophotometer. The essential parts of a spectrophotometer include a light source, a wavelength selection device such as a monochromator or filter, a sample chamber, a light detector, and a readout device, usually also include a computer, which is used for storage and analysis of the spectra. The most useful machines scan the wavelength of the light that is incident on the sample and produce, as output, spectra of absorbance versus wavelength.

Action Spectra

The use of action spectra has been central to the development of our current understanding of photosynthesis. An action spectrum is a graph of the magnitude of the biological effect observed as a function of wavelength. Examples of effects measured by action spectra are oxygen evolution and hormonal growth responses due to the action of phytochrome. Often an action spectrum can identify the chromophore responsible for a particular light-induced phenomenon. Action spectra were instrumental in the discovery of the existence of the two photosystems in O_2-evolving photosynthetic organisms.

Some of the first action spectra were measured by T. W. Engelmann in the late 1800s. Engelmann used a prism to disperse sunlight into a rainbow that was allowed to fall on an aquatic algal filament. A population of O_2-seeking bacteria was introduced into the system. The bacteria congregated in the regions of the filaments that evolved the most O_2.

These were the regions illuminated by blue light and red light, which are strongly absorbed by chlorophyll. Today, action spectra can be measured in room-sized spectrographs in which the scientist enters a huge monochromator and places samples for irradiation in a large area of the room bathed by monochromatic light. But the principle of the experiment is the same as that of Engelmann's experiments.

Schematic diagram of the action spectrum measurements by T. W. Engelmann. Engelmann projected a spectrum of light onto the spiral chloroplast of the filamentous green alga *Spirogyra* and observed that oxygen-seeking bacteria introduced into the system collected in the region of the spectrum where chlorophyll pigments absorb. This action spectrum gave the first indication of the effectiveness of light absorbed by accessory pigments in driving photosynthesis.

Difference Spectra

An important technique in studies of photosynthesis is light-induced difference spectroscopy, which measures changes in absorbance. In this technique, bright light, often called actinic light, is used to illuminate a sample, while a dim beam of light is used to measure the absorbance of the sample at wavelengths other than that of the actinic beam. In this way a difference spectrum is obtained, which represents the changes in the absorption spectrum of the sample induced by illumination with the actinic light. Absorption bands that disappear upon illumination appear as negative peaks; new bands that appear upon illumination appear as positive peaks. Difference spectra give important clues to the identity of molecular species participating in the photoreactions of photosynthesis.

By the use of special flash techniques, it is possible to record the difference spectrum at a given time after flash excitation. Multiple difference spectra recorded at different times after flash excitation can be used to measure the kinetics of the chemical reactions that follow photon excitation of a reaction centre. These techniques can have extraordinary time resolution, in some cases less than a picosecond (10^{-12} s), and have provided great insights into the earliest events in the photosynthetic energy storage process.

Antagonistic Effects of Light on Cytochrome Oxidation

The puzzling red drop and enhancement effects were explained by experiments performed by Louis Duysens of the Netherlands. He suggested that two phtotchemical events were responsible for enhancement effects. One photochemical event produced an oxidation while the other produced a reduction. Chloroplasts contain cytochromes, iron-containing proteins that function as intermediate electron carriers in photosynthesis. Duysens found that when a sample of a red alga was illuminated with long-wavelength light, the cytochrome became mostly oxidized. If light of a shorter wavelength was also present, the effect was partly reversed. These antagonistic effects can be explained by a mechanism involving two photochemical events: one that tended to oxidize the cytochrome and one that tended to reduce it.

We know now that in the red region of the spectrum, one of the photoreactions, known as photosystem I (PS-I), absorbs preferentially far-red light of wavelengths greater than 680 nm, while the second, known as photosystem II (PS-II), absorbs red light of 680 nm well and is driven very poorly by far-red light. This wavelength dependence explains the enhancement effect and the red drop effect. Another difference between the photosystems is that Photosystem I produces a strong reductant, capable of reducing $NADP^+$, and a weak oxidant.

Photosystem II produces a very strong oxidant, capable of oxidizing water, and a weaker reductant than the one produced by photosystem I. This reductant of photosystem II rereduces the oxidant produced by photosystem I, which explains the antagonistic effect.

Structures of Two Bacterial Reaction Centres

In 1984, Hartmut Michel, Johann Deisenhofer, Robert Huber, and coworkers in Munich solved the three-dimensional structure of the reaction centre from the purple photosynthetic bacterium *Rhodopseudomonas viridis*.

This landmark achievement, for which a Nobel prize was awarded in 1988, was the first high-resolution X-ray structure determination for an integral membrane protein and the first structure determination for a reaction centre complex.

The protein part of the complex consists of four separate polypeptides. Two of them, called L and M (for *l*ight and *m*edium mass) bind all of the bacteriochlorophyll, quinone, and carotenoid cofactors of the complex. The structure has a twofold symmetry about an axis perpendicular to the plane of the membrane, hinting at a dimeric nature of the reaction centre. The ten transmembrane portions of the L and M peptides (five from each) are arranged in α helices, and there are almost no charged amino acid residues in the interior of the membrane.

The H (*h*eavy) protein has a single transmembrane helix and is localized mostly on the cytoplasmic side of the membrane. The C (*c*ytochrome) subunit is located in the periplasmic region (the region between the bacterial plasma membrane and the outer membrane). The geometric arrangement of the pigments and the quinones (electron acceptors), with the protein removed. A similar arrangement is found in the reaction centre of another purple photosynthetic bacterium, *Rhodobacter sphaeroides*, except that the C subunit is not present.

Two of the bacteriochlorophyll molecules are in intimate contact with each other and are known as the special pair. This dimer, whose existence was predicted from magnetic-resonance studies, is the photoactive portion of the complex. An electron is transferred from this dimer along the sequence of electron carriers on the right side of the complex. Detailed analysis of these structures, along with analysis of numerous mutants, has revealed many of the principles involved in the energy storage processes that are carried out by all reaction centres.

The bacterial reaction centre structure is thought to be similar in many ways to that found in photosystem II from oxygen-evolving organisms, especially in the electron acceptor portion of the chain. The proteins that make up the core of the bacterial reaction centre are relatively similar in sequence to their photosystem II counterparts, implying an evolutionary relatedness.

Midpoint Potentials and Redox Reactions

Redox reactions, midpoint potentials, and their relationship to the laws of thermodynamics. These concepts are useful for our discussion of electron flow from H_2O to $NADP^+$ and the interactions between the different electron carriers.

The midpoint potential (E_m) is a measure of the tendency of a compound to take electrons from other compounds. A large positive midpoint potential means that the compound is a strong oxidant; a large negative value means that the compound is a strong reductant (in both cases relative to the standard hydrogen electrode).

Equilibrium constants can easily be predicted from midpoint potentials, in the same way that free energies were related to equilibrium constants. Midpoint potentials for many chemical and biochemical reactions have been measured and tabulated. The y-axis on the Z scheme, midpoint potentials of the electron carriers, with negative values higher than positive ones. This choice makes reactions that are spontaneous (releasing free energy) appear "downhill" on the graph.

Knowledge of the midpoint potentials of the various electron carriers is important in establishing the pathway of electron flow in any biochemical electron transport system, such as those found in chloroplasts or mitochondria. Researchers make this measurement usually by carrying out a redox titration.

They adjust, or poise, the sample at a particular redox potential, usually by adding small amounts of oxidants or reductants. Redox mediators, small molecules that permit rapid equilibration between the sample and the electrodes of the measurement system, must be included to ensure that the system is at equilibrium when the measurement is made.

Several measurements are made at a variety of redox potentials. The sample is stirred in a special cell that contains platinum and reference electrodes, and chemical oxidants and reductants are added to adjust the redox potential, which is read with a voltmeter.

We can measure the extent of the redox reaction by following a particular property of the sample, usually absorbance, at each potential. In the example illustrated, the reduced form of the compound has an absorbance that decreases as the compound is oxidized. The fraction of reduced form at each potential is plotted against redox potential, and the midpoint potential (E_m) is determined as the potential at which the compound is half oxidized and half reduced.

Oxygen Evolution

The chemical mechanism of photosynthetic water oxidation is not yet known, although there is a great deal of indirect evidence about the process. If a sample of dark-adapted photosynthetic membrane is exposed to a sequence of very brief, intense flashes, a characteristic pattern of oxygen production is observed. Little or no oxygen is produced on the first two flashes, and maximal oxygen is released on the third flash and every fourth flash thereafter, until eventually the yield per flash damps to a constant value. This remarkable result was first observed by Pierre Joliot in the 1960s.

A schematic model explaining these observations, proposed by Kok and coworkers, has been widely accepted. This model for the photooxidation of water, called the S state mechanism, consists of a series of five states, known as S_0 to S_4, which represent successively more oxidized forms of the water-oxidizing enzyme system, or oxygen-evolving complex. The light flashes advance the system from one S state to the next, until state S_4 is reached. State S_4 produces O_2 without further light input and returns the system to S_0. Occasionally, a centre does not advance to the next S state upon flash excitation, and less frequently, a centre is activated twice by a single flash.

These misses and double hits cause the synchrony achieved by dark adaptation to be lost and the oxygen yield eventually to damp to a constant value. After this steady state has been reached, a complex has the same probability of being in any of the states S_0 to S_3 (S_4 is unstable and occurs only transiently), and the yield of O_2 becomes constant. States S_2 and S_3 decay in the dark, but only as far back as S_1, which is stable in the dark.

Therefore, after adaptation to the dark, approximately three-fourths of the oxygen-evolving complexes appear to be in state S_1 and one-fourth in state S_0. This distribution of states explains why the maximum yield of O_2 is observed after the third of a series of flashes given to dark-adapted chloroplasts.

Photosystem I

The PS-I reaction centre is composed of a multiprotein complex. The reaction centre chlorophyll P700 and about 100 core antenna chlorophylls are bound to two proteins, PsaA and PsaB, with molecular masses in the range of 66 to 70 kDa. PS-I reaction centre complexes have been isolated from several organisms and found to contain the 66 to 70 kDa proteins, along with a variable number of smaller proteins in the range of 4 to 25 kDa. Some of these proteins serve as binding sites for the soluble electron carriers plastocyanin and ferredoxin. The functions of some of the other proteins are not well understood. An 8-kDa protein contains some of the bound iron-sulfur centres that serve as early electron acceptors in photosystem I. The structure of the PS-I complex from pea has been determined to a resolution of 4.4 E, and the positions of many of the chlorophylls and electron transfer components have been located.

In their reduced form, the electron carriers that function in the acceptor region of photosystem I are all extremely strong reducing agents. These reduced species are very unstable and thus difficult to identify. Evidence indicates that one of these early acceptors is a chlorophyll molecule, and another is a quinone species, phylloquinone, also known as vitamin K_1.

Additional electron acceptors include a series of three membrane-associated iron–sulfur proteins, or bound ferredoxins, also known as Fe–S centres Fe–S_X, Fe–S_A, and Fe–S_B. Fe–S centre Fe–S_X is part of the P700-binding protein; centres Fe–S_A and Fe–S_B reside on an 8 kDa protein that is part of the PS-I reaction centre complex. Electrons are transferred through centres Fe–S_A and Fe–S_B to ferredoxin, a small, water-soluble iron–sulfur protein. The membrane-associated flavoprotein ferredoxin–NADP reductase (FNR) reduces $NADP^+$ to NADPH, thus completing the sequence of noncyclic electron transport that begins with the oxidation of water. Electron transport between PSII and PSI is mediated by plastohydroquinone, the cytochrome b_6f complex, and plastocyanin. The cytochrome b_6f complex contains two *b*-type hemes and one *c*-type heme.

A model for the organization of electron carriers in PS-I is shown in Figure below. The P700 dimer is located at the bottom of the structure and two symmetrical arms radiate from P700. Each arm includes an accessory chlorophyll *a* and another chlorophyll molecule tentatively identified as A_0. Other structures include the Fe-S centre

F_X, and two other Fe-S centres, F_1 and F_2. The distance between the carriers is shown at the right.

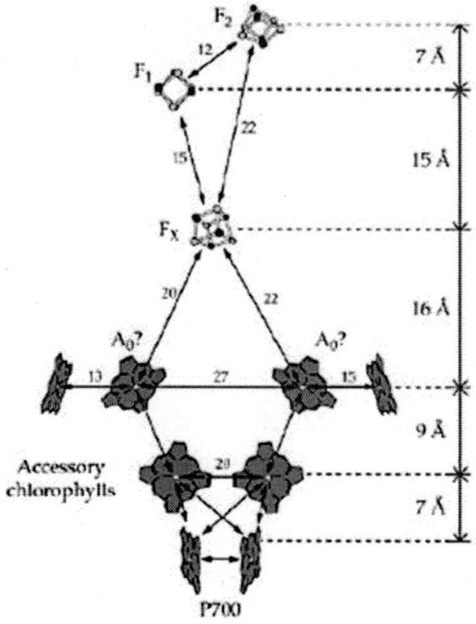

Figure: *A model for the organization of electron carriers in PS-I.*

The PS-I reaction centre appears to have some functional similarity to the reaction centre found in the anaerobic green sulfur bacteria and the heliobacteria. These bacteria contain low-potential Fe–S centres as early electron acceptors and are probably capable of ferredoxin-mediated NAD^+ reduction similar to the $NADP^+$ reduction function of photosystem I. There is almost certainly an evolutionary relationship between these complexes and photosystem I of oxygen-evolving organisms.

ATP Synthase

The chemical mechanism by which ATP is synthesized is not yet understood in detail. However, considerable evidence now supports a mechanism, first proposed by Paul Boyer, in which the principal energy-requiring step is the release of bound ATP from the enzyme.

It has also been proposed that during catalysis a large portion of the CF_1 complex rotates about a bearing consisting of the γ subunit. The γ subunit may act as a camshaft does, rotating alternately against the α and α subunits. The energy of the conformational movements is then translated into phosphoanhydride bond energy.

The thylakoid ATP synthase has two segments: CF_0 is a transmembrane segment, and CF_1 is a hydrophilic segment at the stomal surface. CF_0 traslocates protons from the chloroplast lumen to the catalytic part of the enzyme and CF_1 cataylses the conversion of ADP and inorganic phosphate to ATP.

The chloroplast ATP synthase has nine different subunits. CF_1 is made of two large subunits, α and β, plus three smaller subunits, γ, δ, and ε. Each CF_1 molecule has three copies of the α and β subunits and one copy of the γ, δ and ε subunits. The α and β subunits bind ADP and phosphate and catalyize the phosphorylation of ADP into ATP.

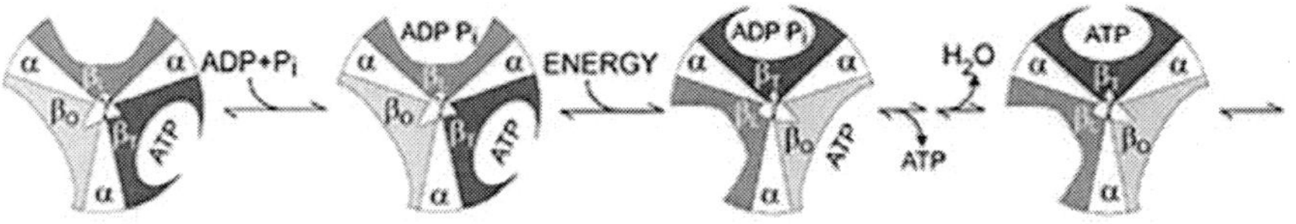

Figure: *The model of Boyer's binding change mechanism. The α and β subunits configure three nucleotide binding sites: O, which provides the early binding site for ADP and inorganic phosphate, L, to which the ADP and inorganic phosphate bind after migrating from O, and T, which tightly binds ATP. Energy ensuing from the movement of protons from the chloroplast lumen to the stroma drives the rotation of the γ subunit of CF_1, and the interconversion of the binding sites and the release of an ATP molecule.*

Studies on the mitochondrial ATP synthase have increased our understanding of how the ATP synthases work. As proposed by Paul Boyer in 1997, ATP synthesis takes place by a binding change mechanism. According to this model, the energy of the proton gradient is used to release a tightly bound form of ATP from a catalytic binding side of the enzyme. As shown in Figure below, CF_1 has three nucleotide binding sites, L, T and O. ADP and inorganic phosphate are postulated to first bind to the O site, and then move to the L site. The T site binds ATP. As protons move from the lumen to the stromal region through the CF_0 channel, energy is released and the γ subunit of CF_1 rotates.

The rotation causes conformation changes in the three nucleotide binding sites, which interconvert, thus changing the afiity of the sites for the nucleotides. As the T site converts into a O site, ATP is released and another cycle starts. Direct observation of the rotation of the γ subunit of CF_1 has provided strong experimental support for the binding change mechanism.

Mode of Action of Some Herbicides

Herbicides of one major class—about half of the commercially important compounds—act by interrupting photosynthetic electron flow. A shows the chemical structure of two of these compounds. The precise sites of action of many of these agents have been found to lie either at the reducing side of photosystem I (for example, in paraquat) or in the quinone acceptor complex in the electron transport chain between the two photosystems.

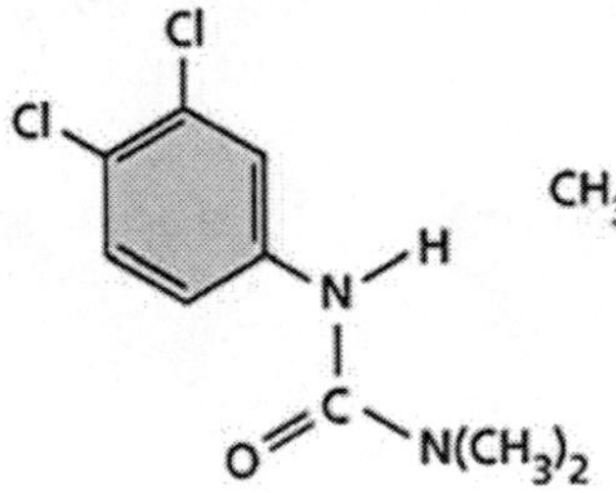

DCMU (diuron)
(Dichlorophenyl-
dimethylurea)

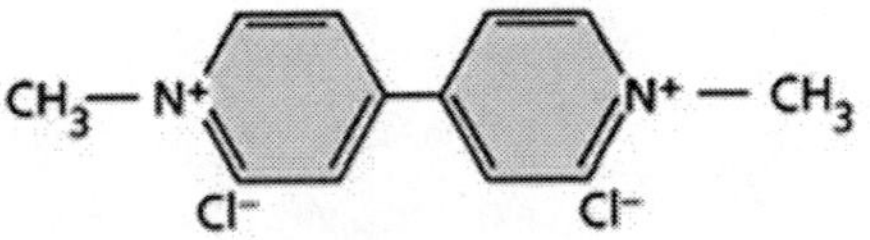

Paraquat
(methyl viologen)

Figure: *The use of herbicides to kill unwanted plants is widespread in modern agriculture. Many different classes of herbicides have been developed, and they act by blocking amino acid, carotenoid, or lipid biosynthesis or by disrupting cell division. Understanding the mode of action of herbicides has been an important tool in research on plant metabolism and has facilitated their application under different agricultural practices (Ashton and Crafts 1981). Here, the chemical structure of two herbicides that block photosynthetic electron flow are shown. DCMU is also known as diuron. Paraquat has acquired public notoriety because of its use on marijuana crops.*

Paraquat acts by intercepting electrons between the bound ferredoxin acceptors and NADP and then reducing oxygen to superoxide (O_2^-). Superoxide is a free radical that reacts nonspecifically with a wide range of molecules in the chloroplast, leading to the rapid loss of chloroplast activity. Lipid molecules in cell membranes are especially sensitive.

The herbicides that act on the quinone acceptor complex compete with plastoquinone for the Q_B binding site. If herbicide is present, it displaces the oxidized form of plastoquinone and occupies the specific binding site for the quinone acceptor, which is thought to lie on the D1 herbicide-binding protein. The herbicide is not able to accept electrons, so the electron is unable to leave Q_A, the first quinone acceptor. Thus, the binding of herbicide effectively blocks electron flow and inhibits photosynthesis. Many herbicides that act in this manner also inhibit electron flow in photosynthetic bacteria that have quinone-type electron acceptor complexes.

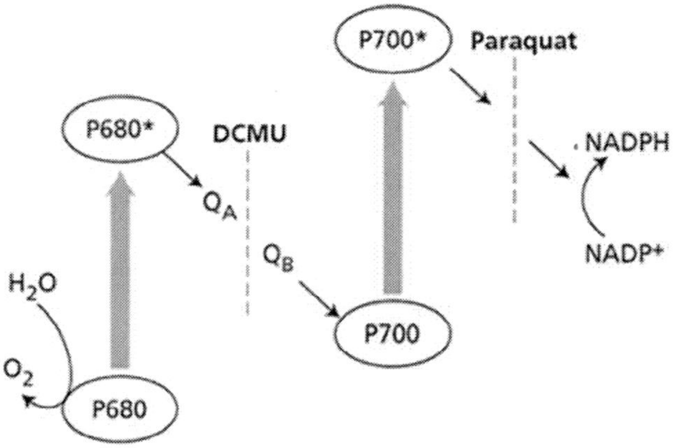

Figure: *Sites of action of herbicides. Many herbicides, such as DCMU, act by blocking electron flow at the quinone acceptors of photosystem II, by competing for the binding site of plastoquinone that is normally occupied by Q_B. Other herbicides, such as paraquat, act by accepting electrons from the early acceptors of photosystem I and then reacting with oxygen to form superoxide, O_2^-, a species that is very damaging to chloroplast components, especially lipids.*

In recent years, herbicide-resistant biotypes of common weeds have appeared in areas where a single type of herbicide has been used continuously for several years. These biotypes can be orders of magnitude more resistant to certain classes of herbicides than the nonresistant plant is. In several cases the resistance factor has been traced to a single amino acid substitution in the D1 protein. This change, presumably in the quinone (and herbicide) binding region of the peptide, lowers the binding affinity of the herbicide, making it much less effective.

The possibility of fine-tuning the herbicide sensitivity of crop plants by making subtle changes in the proteins of the PS-II reaction centre has created a great deal of interest in the agricultural chemical industry. Through biotechnology, it is now possible to make a crop plant that is resistant to a particular herbicide, which can then be applied to control undesirable plants that are not resistant. The success of this approach will depend on whether undesirable side effects of the herbicide-resistant mutations can be controlled and on how rapidly the weeds acquire resistance through natural selection or gene transfer. Thus, it is a race between science and evolutionary change.

A complementary approach is to insert resistance factors to certain pests, such as insects, into the plant via genetic engineering. The plants then produce the toxic species themselves, eliminating the need for insecticides.

Chlorophyll Biosynthesis

In the first phase of chlorophyll biosynthesis, the amino acid glutamic acid is converted to 5-aminolevulinic acid (ALA). This reaction is unusual in that it involves a covalent intermediate in which the glutamic acid is attached to a transfer RNA molecule. This is one of a very small number of examples in biochemistry in which a tRNA is utilized in a process other than protein synthesis. Two molecules of ALA are then condensed to form porphobilinogen (PBG), which ultimately form the pyrrole rings in chlorophyll. The next phase is the assembly of a porphyrin structure from four molecules of PBG. This phase consists of six distinct enzymatic steps, ending with the product protoporphyrin IX.

All the biosynthesis steps up to this point are the same for the synthesis of both chlorophyll and heme. But here the pathway branches, and the fate of the molecule depends on which metal is inserted into the centre of the porphyrin. If magnesium is inserted by an enzyme called magnesium chelatase, then the additional steps needed to convert the molecule into chlorophyll take place; if iron is inserted, the species ultimately becomes heme.

The next phase of the chlorophyll biosynthetic pathway is the formation of the fifth ring (ring E) by cyclization of one of the propionic acid side chains to form protochlorophyllide. The pathway involves the reduction of one of the double bonds in ring D, using NADPH. This process is driven by light in angiosperms and is carried out by an enzyme called protochlorophyllide oxidoreductase (POR). Non-oxygen-evolving photosynthetic bacteria carry out this reaction without light, using a completely different set of enzymes. Cyanobacteria, algae, lower plants, and gymnosperms contain both the light-dependent POR pathway and the light-independent pathway. Seedlings of angiosperms grown in complete darkness lack chlorophyll, because the POR enzyme requires light. These *etiolated* plants very rapidly turn green when exposed to light. The final step in the chlorophyll biosynthetic pathway is the attachment of the phytol tail, which is catalyzed by an enzyme called chlorophyll synthetase. The elucidation of the biosynthetic pathways of chlorophylls and related pigments is a difficult task, in part because many of the enzymes are present in low abundance. Recently, genetic analysis has been used to clarify many aspects of these processes.

Photosynthesis: Carbon Reactions

How the Calvin Cycle Was Elucidated

The elucidation of the Calvin cycle was the result of a series of elegant experiments by Melvin Calvin and his colleagues in the 1950s, for which a Nobel prize was awarded in 1961. They used suspensions of the unicellular eukaryotic green alga *Chlorella* to trace the path of carbon.

The Calvin cycle is also referred to as the reductive pentose phosphate cycle to distinguish the photosynthetic cycle from the oxidative pentose phosphate pathway, with which it shares several enzymes. Yet another designation of the Calvin cycle is photosynthetic carbon reduction cycle.

To elucidate the cycle, the investigators first exposed the algal cells to constant conditions of light and CO_2 to establish steady-state photosynthesis. Next they added radioactive $^{14}CO_2$ for a brief period to label the intermediates of the cycle. They then killed the cells and inactivated their enzymes by plunging the suspension into boiling alcohol. They separated the ^{14}C-labeled compounds from one another and identified them by their positions on two-dimensional paper chromatograms.

In this manner, 3-phosphoglycerate and the various sugar phosphates were identified as intermediates in carbon fixation. By exposing the cells to $^{14}CO_2$ for progressively shorter periods of time, Calvin's group was able to identify 3-phosphoglycerate as the first stable intermediate. Therefore, the other labelled sugar phosphates must be derived from a subsequent reduction of 3-phosphoglycerate.

To deduce the path of carbon, it was necessary to determine the distribution of ^{14}C in each of the labelled sugars. After a brief exposure to $^{14}CO_2$, individual intermediates were isolated and chemically degraded so that the amount of ^{14}C in each carbon atom could be determined. The results showed that 3-phosphoglycerate was initially labelled predominantly in the carboxyl group. This finding suggested that the initial CO_2 acceptor was a two-carbon compound, and it prompted a long and futile search for such a compound.

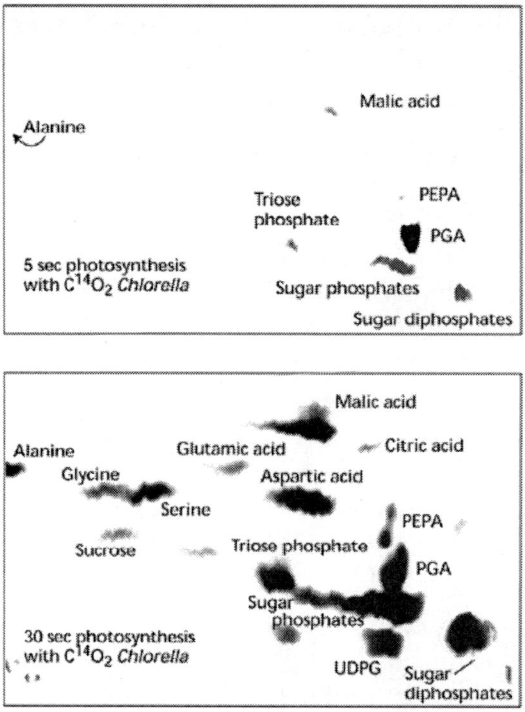

Figure: *Autoradiograms showing the labelling of carbon compounds in the alga Chlorella after exposure to $^{14}CO_2$. The time intervals shown in the figures indicate the length of exposure to the radiolabel. At the indicated time intervals, the reaction was terminated, by plunging of the contents into boiling alcohol. The labelled compounds in the cell homogenates were then separated by paper chromatography. The heavy labelling of 3-phosphoglycerate (PGA) after the shorter exposure indicates that it is the first stable intermediate of the Calvin (reductive pentose phosphate) cycle.*

The subsequent discovery that pentose monophosphates and a pentose bisphosphate (ribulose) participate in the cycle raised the possibility that the initial CO_2 acceptor was a five-carbon compound. This conceptual breakthrough rapidly led to the identification of ribulose-1,5-bisphosphate as the CO_2 acceptor and to the formulation of the complete cycle. The operation of the cycle explained the ^{14}C-labeling pattern of the other intermediates.

To demonstrate conclusively that a metabolic pathway exists, it is necessary to prove that the postulated enzymes catalyze the proposed reactions in the test tube. Ideally, the rates of these enzyme reactions (in vitro) should be equal to or in excess of those observed in the intact cell (in vivo). However, this evidence can be used only to support a proposed pathway. Failure to demonstrate a reaction in vitro does not prove that the reaction does not occur.

All of the reactions of the Calvin cycle have been demonstrated in vitro. With one exception, the in vitro rates of enzymatic activity are well in excess of the maximal observed rates of photosynthesis. That exception is rubisco, which, when assayed under the levels of CO_2 and O_2 in air, shows barely enough in vitro activity to account for the observed rate of photosynthesis in air. Although great progress has been made in increasing activity by providing proper conditions for activation, researchers continue to debate whether rubisco is a rate-limiting step in photosynthesis.

Rubisco: A Model Enzyme for Studying Structure and Function

With the exception of the carbon fixed by a some prokaryotic organisms, most of the carbon fixed on Earth is processed by the Calvin cycle. The concentration of rubisco, the carboxylating enzyme of the Calvin cycle, is generally high. For example, it accounts for 50% or more of the total protein in plant leaves, and its concentration within chloroplasts is extremely high (*ca.* 0.2 g/ml).

Rubisco occurs in two functionally analogous forms, having different structure, distribution, and O_2 sensitivity. Oxygenic phototroph (cyanobacteria, chloroplasts from high plants) and many photosynthetic bacteria contain a form of the enzyme, (form I), made up of eight large (L) catalytic subunits (about 55 kDa each) and eight small (S) subunits (about 14 kDa each), giving a molecular mass of about 560 kDa for the complete protein (L_8S_8). The 8 L subunits of form I are arranged as an octameric core surrounded by two layers of four S subunits, with each layer located on opposite sides of the molecule.

Form II of rubisco is found in some photosynthetic purple nonsulfur bacteria, and it is composed of two L subunits, each 50 kDa. Eukaryotic algal dinoflagellates also use a dimeric enzyme of proteobacterial ancestry. A rubisco-like enzyme lacking catalytic activity has recently been identified in photosynthetic green sulfur bacteria which lack the Calvin cycle. The evidence suggests that the enzyme may function in sulfur oxidation to provide electrons for the photosynthetic electron transport chain. Genome sequences show that the rubisco-like protein also occurs in certain heterotrophic bacteria.

Rubisco Activase

Rubisco catalyzes the incorporation of CO_2 to the C-2 of the enediol form of ribulose-1,5-bisphosphate and the subsequent cleavage of the unstable intermediate to yield two molecules of 3-phosphoglycerate The conversion of the inactive form of rubisco to an active state, which is required for both the carboxylation and oxygenation reactions, requires the carbamylation of a specific lysine residue (in spinach, Lys-201). As a consequence of the binding of the CO_2^- molecule to the lysine residue, the enzyme acquires a triad of anionic residues that provide the binding site for the cofactor Mg^{2+} (in spinach, the other two residues are Asp-203 and Glu-204). At this stage, the incorporation of ribulose-1,5-bisphosphate causes conformational changes in rubisco that allows the enzyme to sequester the sugar bisphosphate from the solvent and, upon the binding of a CO_2 or O_2 molecule, to start the catalytic cycle.

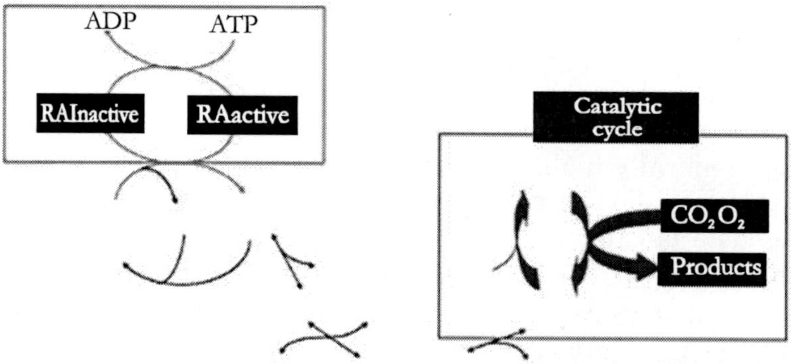

Figure: *Modulation of rubisco by rubisco-activase. The combination or the interaction of rubisco (E) with modulators distributes the enzyme among different states (yellow boxes). The unprotonated form of a specific lysine residue (E-NH_2) reacts with CO_2 releasing a proton and forming the carbamate (E-NH-CO_2^-) that is stabilized by the cofactor Mg^{2+} (E-NH-CO_2^-. Mg^{2+}). At this stage (blue box), rubisco is catalytically competent and thereby binds ribulose-1,5-bisphosphate (RuBP) yielding the ternary complex (E-NH-CO_2^-. Mg^{2+}. RuBP) in which the latter is converted to the enediol form. This entails the incorporation of CO_2 or O_2 in the active site to generate the products of carboxylation or oxygenation, respectively. The high affinity of ribulose-1,5-bisphosphate for free rubisco (E-NH^{3+}) generates the dead-end complex (E-NH^{3+}. RuBP) thereby displacing the above equilibria to a catalytically incompetent state. Rubisco-activase (RA) interacts noncovalently with the binary complex and facilitates the release of ribulose-1,5-bisphosphate (red box) freeing the enzyme for the next carbamylation cycle. This process requires the concurrent hydrolysis of ATP.*

However, the tight binding of ribulose-1,5-bisphosphate to the uncarbamylated enzyme molecules displaces the equilibrium to the dead end complex, rubisco-ribulose-1,5-bisphosphate, so that the rate of reaction declines rapidly until it finally stops. How is this problem solved?

A combination of genetic, physiological and biochemical studies have provided the solution to the long-term riddle of rubisco activation. In vivo, some mutants of *Arabidopsis thaliana* lack light-mediated activation of rubisco and can only grow at high concentrations of CO_2. Surprisingly, the mutation does not impair the response of the enzyme to modulation by CO_2 and Mg^{2+} in vitro. These contradictory results led to the discovery of a protein, rubisco-activase, that uncouples ribulose-1,5-bisphosphate from decarbamylated active sites and, in so doing, promotes the access of CO_2 and Mg^{2+} for the carbamylation of the enzyme.

The importance of this process is that the rate at which rubisco-activase enhances the catalytic capacity of rubisco is linked to the hydrolysis of ATP (50 molecules of ATP hydrolyzed / 1 rubisco site activated). Two isoforms of rubisco activase, (42 and 46 kDa) found in several species differ in the C-terminal region as result of the differential splicing of a common pre-mRNA. Both isoforms catalyze the activation of rubisco and the hydrolysis of ATP but the smaller isoform has higher capacity than the larger isoform for both activities.

Evidence for the fact that the two processes are independent are provided by studies showing that the digestion of native rubisco-activase with trypsin or the modification of its primary structure by site-directed mutagenesis do not elicit the same responses in the stimulation of rubisco and the hydrolysis of ATP. More importantly, the deletion of some residues at the C-region of rubisco-activase doubles the rate of rubisco activation with only a minor effect on the hydrolysis of ATP.

Rubisco activation by the activase requires a mutual recognition of both proteins. Rubisco-activase from tobacco (*Solanaceae*) fails to activate rubisco from spinach or *Chlamydomonas* (non-*Solanaceae*) and, similarly, the rubisco-activase from spinach or *Chlamydomonas* fails to stimulate rubisco from tobacco. However, the substitution of some residues on the surface of the large subunit of *Chlamydomonas* rubisco shifts the specificity from non-*Solanaceae* to *Solanaceae* rubisco-activase.

These experiments underscore the contribution of specific amino acids to protein-protein interaction and provide experimental evidence for a recognition step separated from the subsequent activation process.

Three Variations of C_4 Metabolism

There are three variations of the basic C_4 pathway. The variations differ principally in the C_4 acid transported into the bundle sheath cells (malate or aspartate) and in the manner of decarboxylation.

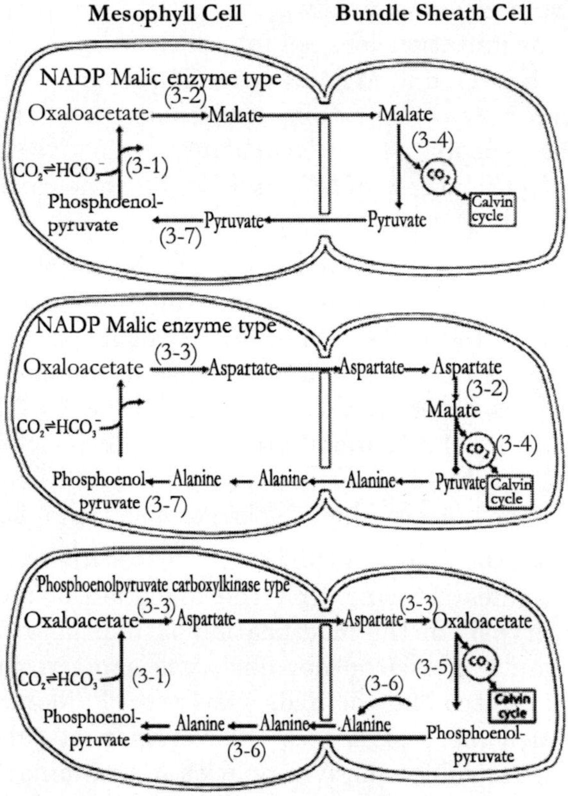

Figure: *The three variants of the C_4 photosynthetic carbon cycle. The variants differ principally in (1) the nature of the four-carbon acid (malate or aspartate) transported into the bundle sheath cell and of the three-carbon acid (pyruvate or alanine) returned to the mesophyll cell and (2) the nature of the enzyme that catalyzes the decarboxylation step in the bundle sheath cell. The three variants are named after the enzymes that catalyze the decarboxylation reactions. Representatives of each variant include maize, crabgrass, sugarcane, sorghum (NADP malic enzyme); pigweed, millet (NAD malic enzyme); guinea grass (phosphoenolpyruvate carboxykinase).*

The variations are named after the enzymes that catalyze the respective decarboxylation reactions: NADP-dependent malic enzyme (NADP-ME) found in chloroplasts; NAD-dependent malic enzyme (NAD-ME) in mitochondria; and phosphoenolpyruvate carboxykinase (PEP-CK) in the cytosol. The discussion in the text traces the route of carbon for the NADP-ME type, such as sugarcane or corn.

The Carbon Path

The primary carboxylation reaction, catalyzed by PEP carboxylase, which is common to all three variants, occurs in the cytosol of the mesophyll cells and uses HCO_3^- rather than CO_2 as a substrate. In the NADP-ME variant, oxaloacetate is rapidly reduced to malate in the mesophyll chloroplasts by NADPH using NADP:malate dehydrogenase. The malate formed enters the chloroplast of the bundle sheath cell and there undergoes oxidative decarboxylation, yielding pyruvate. The CO_2 released within the bundle sheath cells is converted to carbohydrate by the Calvin cycle.

The residual C_3 acid is transported back to the mesophyll as pyruvate and converted to phosphoenolpyruvate in the mesophyll chloroplast. This reaction takes place in the mesophyll chloroplast of all three C_4 variants. The cycle then starts again with carboxylation of the phosphoenolpyruvate to produce oxaloacetate.

Another process localized in the mesophyll chloroplast of all three C_4 variants is the reduction of 3-phosphoglycerate to triose phosphates, which occurs within the Calvin cycle proper in C_3 chloroplasts. In contrast, all C_4 variants export 3-phosphoglycerate from the bundle sheath for reduction in the mesophyll chloroplast. After reduction, the triose phosphates are transported back to the bundle sheath chloroplasts, for use in the operation of the Calvin cycle there.

One possible reason for this remarkable split of the Calvin cycle between the bundle sheath and the mesophyll chloroplast is to restrict electron transport activity and oxygen evolution in the bundle sheath, and thus avoid oxygen levels that would favour photorespiration. The transport of metabolites of the C_4 pathway is accompanied by movements of protons and other ionic species in order to maintain pH and charge balance.

Photorespiration in CAM Plants

Many terrestrial plants, adapted to hot and arid regions where water is scarce, are stressed by the severe effects of full sunlight, particularly drought and high temperature. To avoid high rates of water transpiration, the stomata of these species are tightly closed during the hot diurnal hours and perform most of their gas exchange with the surrounding atmosphere at night, when the air is relatively cool and humid. On the other hand, the vital process of photosynthesis in which CO_2 is incorporated into carbohydrates takes place during the day. To cope with the lack of an efficient diurnal influx of atmospheric CO_2, these species have evolved a different photosynthetic mechanism, called "crassulacean acid metabolism" (CAM).

This unusual process takes place within single leaf cells in which the uptake and subsequent assimilation of atmospheric CO_2 involves three compartments: cytosol, vacuole, and chloroplast. In general, CAM plants are common among desert succulents (e.g., *Cactaceae*) exposed to daytime high temperatures and low relative humidity, or among tropical forest epiphytes (*Bromeliaceae*) which are exposed to physiological drought under humid conditions because of their aerial growth habit. Remarkably, CAM also occurs in aquatic species (*Isoetaceae*) found worldwide in marshy and shallow habitats that exhibit a diurnal limitation of CO_2.

During the cool night-time hours, the open stomata facilitate the uptake of atmospheric CO_2 which diffuses into the cell cytosol, generating HCO_3^-. At this stage, phosphoenolpyruvate-carboxylase (PEPCase) catalyzes the incorporation of HCO_3^- into phosphoenolpyruvate, yielding oxaloacetic acid and related four-carbon organic acids (particularly malic acid), that accumulate in the vacuole. During the hot hours of the day, the stomatal closure characteristic of terrestrial CAM plants reduces water loss but causes an internal CO_2 shortage. To overcome this deficiency, the vacuole releases malic acid into the cytosol where the action of decarboxylating enzymes, NADP–malic enzyme or PEP-carboxykinase, produces pyruvate and CO_2. Inside the stroma of chloroplasts, Rubisco incorporates the released CO_2 into 3-phosphoglycerate which, in concert with the ATP and NADPH synthesized in the light via photophosphorylation, initiates a series of Benson–Calvin cycle reactions that result in the accumulation of starch.

The temporal separation of the nocturnal formation and storage of malic acid, and the diurnal decarboxylation and the accompanying accumulation of plastid starch, proceeds over the course of a 24-hour photoperiod. On this temporal basis, CAM consists of four metabolic phases that encompass the modulation of C_4 (PEPCase) and C_3 (Rubisco) carboxylation within the same cellular environment (Figure). The nocturnal uptake of atmospheric CO_2 via PEPCase to form C_4 acids and the diurnal decarboxylation of malic acid to elicit an elevated internal concentration of CO_2 are interspersed with transitional periods of net CO_2 uptake in early morning and in late afternoon, when both the PEPCase-and the Rubisco-mediated carboxylations contribute to the assimilation of CO_2. In this circadian rhythm, the proportion of CO_2 taken up via PEPCase at night or via Rubisco during the day is controlled by:

- stomatal function
- fluctuations in the content of organic acids and storage carbohydrates
- the activity of enzymes involved in the primary (PEPCase) and the secondary (Rubisco) carboxylations, and in the decarboxylation of organic acids (e.g., malic enzyme or PEP-carboxykinase)

In diurnal phase III, the closed stomata preclude gas exchange with the external atmosphere, but photosynthesis and other vital processes continue. At this stage, the internal levels of CO_2 fall precipitously due to an active Benson–Calvin cycle; however, this decline is offset by mitochondrial respiration. The closure of the stomata also leads to the accumulation of O_2 in leaf cells because the rate of photosynthetic O_2 evolution is higher than mitochondrial respiration and other O_2-consuming reactions. At this stage, the oxygenase activity of Rubisco activates the photorespiratory cycle by which the Benson–Calvin cycle recovers 75% of the carbon diverted to glycolate, with the remaining 25% released as CO_2 by the mitochondrial glycine decarboxylase complex.

Thus, the closed stomata provide an adaptive advantage that minimizes water loss from the leaves by transpiration, and keeps the concentration of stromal CO_2 adequate for carbon assimilation. However, the concurrent accumulation of O_2 is a major adverse consequence because, as a powerful oxidant, O_2 reacts with other molecules and generates harmful free radicals (e.g., superoxide, hydroxyl ion) which may initiate uncontrolled adverse reactions on DNA, proteins, and lipids.

In this scenario, photorespiration helps alleviate the undesirable consequences of O_2 accumulation by consuming both stromal and peroxisomal O_2. The release of one mole CO_2 in a single round of the photorespiration cycle requires the successive utilization of: (a) two moles O_2 by the oxygenase activity of Rubisco in the chloroplast stroma, and (b) one mole O_2 for the concerted action of glycolate oxidase and catalase in the small oxygen cycle in the peroxisome. The photorespiratory cycle driven by ribulose-1,5-bisphosphate {2 Ribulose-1,5-bisphosphate + 3 O_2 + H_2O + ATP → 3 3-phosphoglycerate + CO_2 + 2 P_i + ADP} requires substantially more O_2 than mitochondrial respiration operating with carbohydrates {$C_6H_{12}O_6$ (e.g., glucose) + 6 O_2 → 6 CO_2 + 6 H_2O}. Thus, three versus one moles O_2 are consumed, respectively, per mole CO_2 released. It is therefore evident that the beneficial consequence of CAM photorespiration in Phase III involves not only the avoidance of carbon loss due to the formation of glycolate, but also the amelioration of oxidative stress due to removal of reactive oxygen species. This mitigation of oxidative stress has been proposed to be a major driving force in the evolution of CAM.

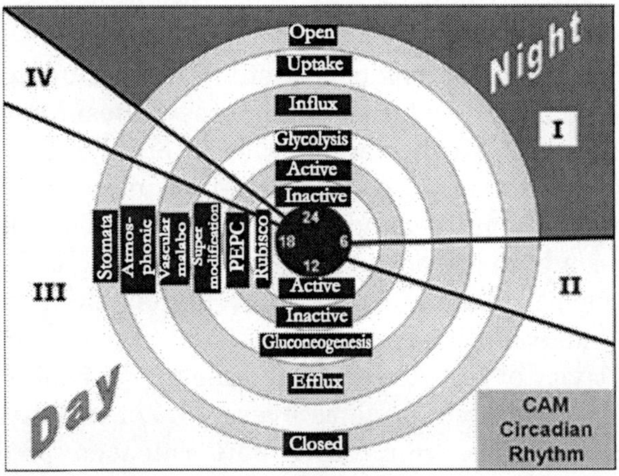

***Figure:** CAM Circadian rhythm. Metabolite fluxes [atmospheric CO_2, vacuolar malate, sugar mobilization to form PEP] and enzyme activities [PEPCase, Rubisco] are indicated on the four major temporal phases of CAM [I, II, III, IV]. Phase I comprises the nocturnal PEPCase-driven uptake of atmospheric CO_2 and malic acid accumulation. Phase II is the transitional phase wherein an accelerated burst of CO_2 uptake takes place due to both the PEPCase-mediated and the Rubisco-mediated fixation of CO_2. During diurnal Phase III, the uptake of atmospheric CO_2 is diminished and the decarboxylation of the vacuolar malic acid generates the stromal concentration of CO_2 needed for Rubisco-mediated CO_2 fixation via the Benson–Calvin cycle. Phase IV may include CO_2 uptake that is fixed directly by Rubisco.*

Fructans

Although starch is the dominant carbohydrate reserve in many species, it may be almost absent in the storage organs of some plants. Fructans are less familiar reserve carbohydrates but occur in 15% of flowering species that mainly belong to the Asteraceae, Campanulaceae and Boraginaceae (dicots), and Poaceae and Liliaceae (monocots).

This type of polysaccharide, whose basic unit is fructose, was found in 1804 in a hot water extract from *Inula helenium* and later named "inulin." Fructans can replace starch as a reserve carbohydrate in some plants, though in many species both polysaccharides coexist. Moreover, the carbohydrate reserve substances vary among plant species and even within a single plant species throughout the seasons and under various climatic conditions. In contrast to starch—which is stored in plastids—fructans appear in vacuoles of different organs: for example, in bulbs (onion), tubers (Jerusalem artichoke, dahlia), taproots (chicory), leaves, and in stems (wheat, barley).

Chemically, fructans are distinguished on the basis of the glycosidic bond that links fructose residues to each other. The biosynthesis of these polysaccharides starts with the incorporation of a fructose moeity to one of the three primary hydroxyl groups of vacuolar sucrose—that is, carbon-1 and carbon-6 in the fructose moeity, and carbon-6 in the glucose moeity (Figure above). This reaction, catalyzed by the enzyme sucrose:sucrose 1-fructosyl transferase, implies a transglycosylation in which one molecule of sucrose transfers the fructosyl moeity to another. Thus, the addition of β-D-2,1-or β-D-2,6-linked fructofuranosyl units to the fructose moeity of sucrose yields 1-kestose and 6-kestose, respectively, while the addition of a β-D-2,6-linked fructofuranosyl unit to the glucose moeity of sucrose produces neo-kestose.

Further elongation of the ketose-type trisaccharide proceeds by the action of specific enzymes that transfer, also via transglycosylation, a fructosyl unit from a fructan molecule to positions 1 or 6 in the second fructose moeity of the trisaccharide. Thus, fructan:fructan fructosyl transferases catalyze the elongation of kestoses yielding the linear β-D-2,1-bonds of inulins, mainly in dicots, and the β-D-2,6-linkages in levans of monocots and bacteria.

$$\text{GFF (kestose)} + \text{G(F)}_m \rightarrow \text{G(F)}_3 + \text{G(F)}_{m-1}$$

While up to 100,000 fructose moeities can be linked in a single molecule of bacterial fructan, fructosyl units range from 10 to 200 in plant counterparts. Fructans generally are present as a continuum of oligofructans that differ from each other in one fructose residue: $G(F)_3$, $G(F)_4$, $G(F)_5$, $G(F)_6$,..., $G(F)_n$. This feature enables fructans to participate in the regulation of the cellular osmotic potential.

In addition to supplying hexose units for demands of cellular energy, the breakdown of the endogenous polysaccharide elevates the concentration of lower members in the oligofructans series, quickly increasing in consequence the osmotic pressure.

Two types of fructan hydrolases cleave the glycosidic bond in fructans: exohydrolases that release terminal fructosyl residues while endohydrolases cleave at random within the fructan chains. Apparently, plants contain only the former type while bacteria and fungi contain both kinds of enzymes.

The recent finding of fructan exohydrolases in nonfructan plants not only suggests that these enzymes are additionally involved in other important functions—perhaps defence and signalling—but also highlights the necessity of functional characterization of these enzymes (Van den Ende et al. 2004).

Plastidic Phosphate Translocators

Chloroplasts are metabolic powerhouses within leaf cells. They are not only the main site of carbon, nitrogen, and sulfur assimilation but also share metabolic pathways—such as the photorespiratory carbon cycle—with other plant cell compartments. A continuous exchange of metabolites and ions with the cytosol is necessary for playing these roles.

In all vascular plants, plastids are surrounded by an envelope that restricts the nonspecific diffusion of polar molecules. Thus, a pair of concentric membranes modulate the exchange of metabolites between the stromal interior and the cytosol. Although many pore-forming proteins that are substrate-specific were characterized in the outer membrane, the inner membrane is considered the main permeability barrier between the cytosol and the chloroplast. All transporters of the inner membrane are nuclear-encoded membrane proteins whose precursors are synthesized in the cytosol and post-translationally imported into the inner membrane.

Chloroplast transporters collectively grouped as phosphate translocators catalyze a strict 1:1 counter-exchange (antiport) of orthophosphate with triose phosphates, pentose phosphates, and hexose phosphates. The Arabidopsis genome contains six genes that encode four classes of functional plastidial phosphate translocators (Table): single copies of the triose phosphate/phosphate translocator (TPT) and the phosphoenolpyruvate/phosphate translocators (PPT); and two copies each of the glucose 6-phosphate/phosphate translocators (GPT) and the xylulose 5-phosphate/phosphate translocator (XPT).

Table: *Plastidic Phosphate Translocators*

Translocator	***MATDB entry***	***Compounds transported***
TPT	TPT (At5g46110)	triose phosphates, 3-phosphoglycerate
PPT	PPT1 (At5g54800),	
	PPT2 (At3g01550)	phosphoenolpyruvate, 2-phosphoglycerate
GPT	GPT1 (At5g33320),	
	GPT2 (At1g61800)	glucose 6-phosphate, triose phosphates
XPT	(At5g17630)	xylulose 5-phosphate, triose phosphates

The Triose Phosphate/Phosphate Translocator and the Allocation of Carbon in Plants

Triose phosphates, generated by the Calvin cycle at the expense of photosynthetically generated ATP and NADPH, flow to the cytosol across the chloroplast envelope.

In turn, the orthophosphate released in the cytosol from biosynthetic processes is transported back into chloroplasts to replenish the ATP necessary for sustaining the assimilation of CO_2 and photosynthetic electron transport.

Both the efflux of triose phosphates and the influx of orthophosphate are driven by a dimer composed of two identical subunits—the chloroplastic TPT. This membrane protein functions as an antiport system that exchanges phosphorylated—mainly three-, but also admits five-or six-carbon compounds—for orthophosphate.

Hence, the chloroplast TPT exports the fixed carbon for the cytosolic synthesis of sucrose, the main product allocated to heterotrophic plant organs for further metabolism or conversion into storage products (e.g., starch, fructans). Therefore, the chloroplast TPT represents *the diurnal path for carbon export.*

Mutants defective in chloroplast TPT rely on different paths to transfer the photoassimilate from the chloroplast to the cytosol and, in so doing, compensate for the deficiency in TPT activity.

Therefore, knockout mutations of this translocator increase the allocation of recently fixed carbon into transitory starch, followed by the degradation of the starch in the light and the subsequent export of neutral sugars to the cytosol. On the other hand, the simultaneous knockout of chloroplast TPT and starch synthesis drastically impairs plant growth.

The Phosphoenolpyruvate/Phosphate Translocator is Essential for the Chloroplast Metabolism

PPT in the inner chloroplast membrane also links the stromal metabolism of all plants with the surrounding cytosol. C_4-plants of the NADP-malic enzyme type (e.g., maize) under active photosynthesis produce pyruvate in bundle sheath cells. This three-carbon compound returns to mesophyll cells for the conversion to phosphoenolpyruvate by the chloroplast-localized pyruvate-phosphate dikinase (Figure below). Subsequently, the phosphoenolpyruvate flows from the stroma to the cytosol for the primary carbon fixation via PEPCase. To ensure high rates of phosphoenolpyruvate efflux for the proper functioning of this cycle, the envelope of mesophyll chloroplasts bears the PPT, which concurrently drives the chloroplast uptake of orthophosphate to restore the lost phosphorus.

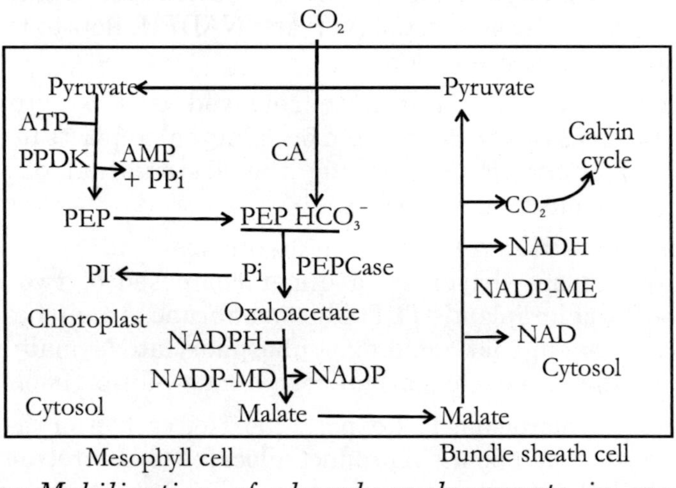

Figure: *Mobilization of phosphoenolpyruvate in mesophyll chloroplasts of C_4plants of the NAD-malic enzyme type. The pyruvate produced by the malic enzyme (ME) in bundle sheath cells flows to mesophyll cells for conversion to phosphoenolpyruvate (PEP) by the pyruvate-phosphate dikinase (PPDK) localized at the chloroplast stroma. The efflux of PEP to the cytosol through the phosphoenolpyruvate/ phosphate translocator (PPT) restores the acceptor of HCO_{3^-}required by the phosphoenolpyruvate carboxylase (PEPCase) for the primary carbon fixation. CA: carbonic anhydrase; NADP-MD: NADP-dependent malate dehydrogenase.*

On the other hand, CAM plants of the malic enzyme-type decarboxylate during the day the malate accumulated at night. Again, two successive processes account for the production of phosphoenolpyruvate, the primary acceptor of CO_2. First, the pyruvate formed in decarboxylation of the malate enters into the chloroplast stroma for its transformation to phosphoenolpyruvate catalyzed by the pyruvate-phosphate dikinase. Second, the phosphoenolpyruvate leaves the chloroplast in exchange for orthophosphate via the PPT.

Knowledge on the function of the PPT in C_3 plants or nongreen tissues of C_4 plants is scarce. Apparently, the uptake—rather than the export—of phosphoenolpyruvate is the primary function of PPT in these tissues. Chloroplasts and most nongreen plastids are devoid of phosphoglyceromutase and enolase and, consequently, are unable to convert 3-phosphoglycerate to phosphoenolpyruvate.

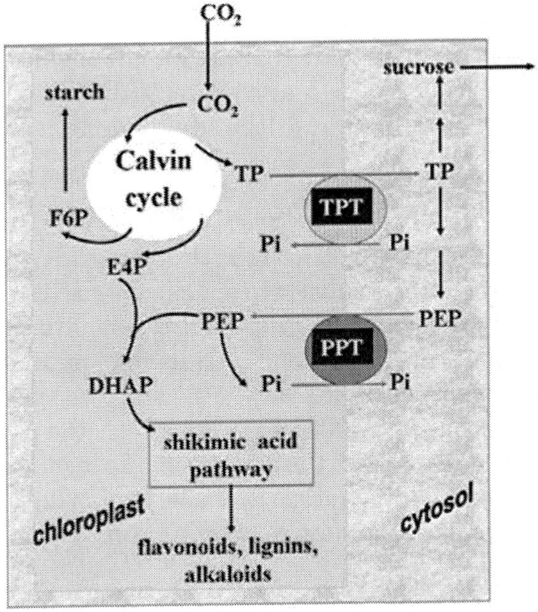

Figure: *Mobilization of phosphoenolpyruvate in chloroplasts of C_3 plants. Under active photosynthesis, TPT furnish the cytosol with carbon skeletons for the biosynthesis of phosphoenolpyruvate. Upon the release of orthophosphate, the uptake of the cytosolic phosphoenolpyruvate, which combines in the stroma with erythrose 4-phosphate, triggers the formation of 3-deoxy-D-arabino-heptulosonic acid-7-phosphate (DHAP). DHAP is a scaffold for the biosynthesis of shikimic acid-5-phosphate, which in turn starts the pathway leading to the aromatic amino acids tyrosine and phenylalanine, the precursors of flavonoids, lignins, and alkaloids.*

However, phosphoenolpyruvate combines with erythrose 4-phosphate, an intermediate of the pentose-phosphate pathway, for the synthesis of aromatic compounds through the shikimic acid pathway (Figure below). Thus, the provision of cytosolic phosphoenolpyruvate to the stroma via the PPT would be essential for the biosynthesis of flavonoids, lignins, alkaloids.

The Glucose 6-Phosphate/Phosphate Translocator Functions in Nongreen Tissues and is Essential for Plant Development

The genome of Arabidopsis harbours two paralogous GPT genes, *AtGPT1* and *AtGPT2*, that, apparently, function in plastids of nongreen tissues for the import of glucose 6-phosphate. Two transgenic lines, bearing T-DNA insertions in the *GPT1* gene, exhibited not only impaired pollen and ovule development but also were lethal in the homozygous state. At variance, the disruption of the *GPT2* genes have no effect on growth and development. On the basis of drastic gametogenesis defects, it appears that the import of glucose 6-phosphate into nongreen plastids by the translocator AtGPT1 is crucial for the maturation of the pollen and the development of the female gametophyte.

The Xylulose 5-Phosphate/Phosphate Translocator

The XPT constitutes the fourth subfamily of phosphate translocators that exhibits substrate specificity similar to the GPT but lacks the capacity to transport glucose 6-phosphate. In Arabidopsis, the lack of cytosolic transketolase and transaldolase halts the cytosolic oxidative pentose phosphate pathway at the stage of pentose phosphates (Kruger and von Schaewen 2003). In this context, the putative function of the XPT would be to furnish the stromal pentose phosphate pathway with cytosolic five-carbon skeletons in the form of xylulose 5-phosphate, especially under conditions that require its removal from the cytosol.

Photosynthesis: Physiological and Ecological Considerations

Working with Light

Three light properties are especially important when working with light: amount, direction, and spectral quality. The first two parameters, amount and direction, are important with respect to the geometry of the part of the plant that intercepts the light. Is the plant part flat or cylindrical? For a flat leaf, a planar light sensor is the most appropriate, and the amount of energy that falls on a flat sensor of known area per unit time is quantified as irradiance. Units can be expressed in terms of energy, such as watts per square meter (W m^{-2}). Time (seconds) is contained within the term watt: 1 W = 1 joule (J)s^{-1}. The energy of a photon depends on its frequency, as expressed by Planck's law.

When considered as a wave, light has a wavelength and a frequency. Light can also be thought of as a stream of particles, photons, or quanta. In this case, units can be expressed in moles per square meter per second (mol m^{-2} s^{-1}), where "moles" refers to the number of photons (1 mol of light = 6.02×10^{23} photons, Avogadro's number). This measure is called photon irradiance.

Quanta and energy units can be interconverted, provided that the wavelength of the light is known. The energy of a photon is related to its wavelength as follows:

$$E = \frac{hc}{\lambda}$$

where *c* is the speed of light (3×10^{8} m s^{-1}), *h* is Planck's constant (6.63×10^{-34} J s), and λ is the wavelength of light, usually expressed in nm (1 nm = 10^{-9} m). We can solve for the *hë* part of the equation, and we obtain $1{,}988 \times 10^{-16}$, and write this equation as:

$$E = \frac{1{,}988 \times 10^{-16}\,J}{\lambda}$$

where λ is expressed in nanometers. From this equation we can see that a photon at 400 nm, which is in the blue region of the spectrum, has twice the energy of a photon at 800 nm, from the infrared region of the spectrum. A photon of 400 nm light contains 4.97×10^{-19} J. On the other hand, the 800 nm photon contains 2.48×10^{-19} J. Stated differently, the higher the wavelength of a photon, the lower its energy, as indicated by the larger denominator in the equation.

Now, suppose we have 3 μmol of 400 nm light falling on 1 m^2 every second, or a photon irradiance of 3 μmol m^{-2} s^{-1}. This quantity is approximately the amount of blue light at 400 nm that strikes the surface of Earth on a sunny day. If we want to convert photon irradiance to energy irradiance, we must first convert micromoles to moles (1 μmol = 10^{-6} mol): (3 μmol m^{-2} s^{-1}) × (10^{-6} mol μmol^{-1}) = 3×10^{-6} mol m^{-2} s^{-1} of photons, or quanta. Calculating the number of quanta from Avogadro's number gives: (3×10^{-6} mol quanta m^{-2} s^{-1}) × (6.02×10^{23} quanta mol^{-1}) = 1.8×10^{18} quanta m^{-2} s^{-1}. As we already calculated, each photon, or quantum, of 400 nm light contains 4.97×10^{-19} J. Thus (4.97×10^{-19} J $quantum^{-1}$) × (1.8×10^{18} quanta m^{-2} s^{-1}) = 0.9 J s^{-1} m^{-2}. Since 1 J s^{-1} = 1 W, we have an irradiance of 0.9 W m^{-2}.

Direction of light. Turning our attention to the direction of light, light can strike a flat surface directly from above or it can strike the surface obliquely. When light deviates from perpendicular, irradiance is proportional to the cosine of the angle at which the light rays hit the sensor.

Thus, irradiance is maximal when light strikes a surface directly from above, and it decreases as light becomes more oblique—similar to the situation with a typical leaf. Sensors that correct for the angle of incidence of light are said to be *cosine corrected.*

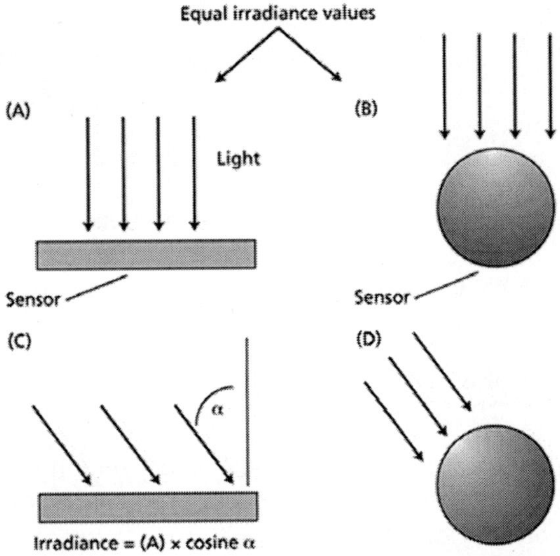

Figure: *Irradiance and fluence rate. Equivalent amounts of collimated light strike a flat irradiance-type sensor (A) and a spherical sensor (B) that measure fluence rate. With collimated light, A and B will give the same light readings. When the light direction is changed 45°, the spherical sensor (D) will measure the same quantity as in B. In contrast, the flat irradiance sensor (C) will measure an amount equivalent to the irradiance in A multiplied by cosine of the angle α in C. (After Björn and Vogelmann 1994.)*

There are many examples in nature in which the light-intercepting object is not flat (e.g., complex shoots, whole plants, chloroplasts). In addition, in some situations light can come from many directions simultaneously (e.g., direct light from the sun plus the light that is reflected upward from sand, soil, or snow). In these situations it makes more sense to measure light with a spherical sensor that measures light omnidirectionally (from all directions).

When the amount of light is measured by this omnidirectional measurement, the type of measurement is called fluence rate (Rupert and Letarjet 1978), and the measured amount of light can be expressed in watts per square meter (W m^{-2}) or moles per square meter per second (mol m^{-2} s^{-1}). It is clear from the units whether light is being measured as energy (W) or as photons (mol).

In contrast to a flat sensor, a spherical sensor is equally sensitive to light from all directions. Depending on whether the light is collimated (rays are parallel) or diffuse (rays travel in random directions), values for fluence rate versus irradiance can be quite different from one another. They are equivalent only under special conditions. The flat sensor measurement of photosynthetically active radiation (PAR, 400 to 700 nm) may also be expressed on the basis of energy (W m^{-2}) or quanta (mol m^{-2} s^{-1}) (McCree 1981). It is important to note that PAR is an irradiance-type measurement. In research on photosynthesis, when PAR is expressed on a quantum basis, it is often given the special term photosynthetic photon flux density (PPFD). However, it has been suggested that the term "density" be discontinued because within the International System of Units (Système Internationale d'Unités, or SI units) "density" can mean area or volume. Moreover, area is contained within the term flux. PPFD has in some cases been shortened to PPF, but it is not clear whether this abbreviation represents an irradiance-type or a spherical measurement.

When light moves through water such as lakes or oceans, it scatters in such a way that its *direction* of travel is randomized. As a result, aquatic photosynthetic organisms can receive light simultaneously from above, below, and the sides. In oceanography and limnology, the light that travels from the water surface toward the bottom is called *downwelling irradiance*, and the light that travels in the reverse direction is called *upwelling irradiance*.

Before the advent of specialized light meters suitable for plant physiology, light was measured in lux or foot-candles. These measurements are based on the perception of light by the human eye, which is maximally sensitive to light within the green region of the spectrum, at 555 nm. Sensitivity falls off on both sides of this wavelength and approaches zero within the blue and red, wavelength regions that are important for photosynthesis. Thus, lux and foot-candles are of little use in plant physiology. These units can be converted to the units of irradiance and fluence rate, but only if detailed knowledge about the spectral quality of the light is available, something that can be obtained only by use of an instrument called a *spectroradiometer*, which measures the amount of light at each wavelength.

In summary, when choosing how to quantify light, it is important to match sensor geometry and spectral response with that of the plant. Flat, cosine-corrected sensors are ideally suited to measure the amount of light that strikes the surface of a leaf; spherical sensors are more appropriate in other situations, such as when studying a chloroplast suspension or a branch from a tree.

How much light is there on a sunny day and what is the relationship between PAR irradiance and PAR fluence rate? Under direct sunlight, PAR irradiance and fluence rate are both about 2000 μmol m^{-2} s^{-1}, though higher values can be measured at high altitudes. The corresponding value in energy units is about 400 W m^{-2}. When light is completely diffuse, irradiance is only 0.25 times the fluence rate.

Heat Dissipation from Leaves: The Bowen Ratio

The heat load on a leaf exposed to full sunlight is very high. In fact, a leaf with an effective thickness of water of 300 μm would warm up by 100°C every minute if all available solar energy were absorbed and no heat was lost. However, this enormous heat load is dissipated by the emission of long-wave radiation, by sensible (or perceptible) heat loss, and by evaporative (or latent) heat loss.

Air circulation around the leaf removes heat from the leaf surfaces if the temperature of the leaf is higher than that of the air; this phenomenon is called sensible heat loss.

Evaporative heat loss occurs because the evaporation of water requires energy. Thus, as water evaporates from a leaf it withdraws heat from the leaf and cools it. The human body is cooled by the same principle, through perspiration.

Sensible heat loss and evaporative heat loss are the most important processes in the regulation of leaf temperature, and the ratio of the two is called the Bowen ratio:

$$\text{Bowen ratio} = \frac{\text{sensible heat loss}}{\text{evaporative heat loss}}$$

This concept was developed by Ira S. Bowen (1898–1978), an American astrophysicist. When the evaporation rate is low, because water supply is limited, the Bowen ratio tends to be high. Thus, the Bowen ratio is about 10 for deserts, 2-6 for semi-arid regions, 0.4 to 0.8 for temperate forests and grasslands, 0.2 for tropical rain forests and 0.1 for tropical oceans

In well-watered crops, transpiration, and hence water evaporation from the leaf, is high, so the Bowen ratio is low. On the other hand, in some cacti, stomata closure prevents evaporative cooling; all the heat is dissipated by sensible heat loss, and the Bowen ratio is infinite. Plants with very high Bowen ratios conserve water but have to endure very high leaf temperatures in order to maintain a sufficient temperature gradient between the leaf and the air. Slow growth is usually correlated with these adaptations.

Very low Bowen ratios can be measured in a lawn on a relatively still day. In these conditions there is no sensible heat loss, because the air around the leaf is at the same temperature as the leaf; the Bowen ratio therefore approaches zero, and heat dissipation is due mostly to evaporative heat loss. In other cases, such as cotton leaves in the afternoon, water loss from stomatal transpiration cools the leaf below the air temperature by evaporative heat loss. In that case, there is sensible heat gain rather than heat loss, and the Bowen ratio becomes negative.

One can calculate the evapotranspiration rate for an entire canopy using measurements of the Bowen ratio, net incident radiation, the heat loss from the soil, and the gradients in temperature and water vapour concentration above the canopy.

The Geographic Distributions of C_3 and C_4 Plants

Among the 15,000+ species with C_4 photosynthesis, it is most common in grasses and sedges, less common in herbs and shrubs, and not found in trees (with a single Hawaiian tree exception, *Euphorbia forbesii*). Climate is a major factor influencing the natural distributions of C_3 plants and C_4 grasses.

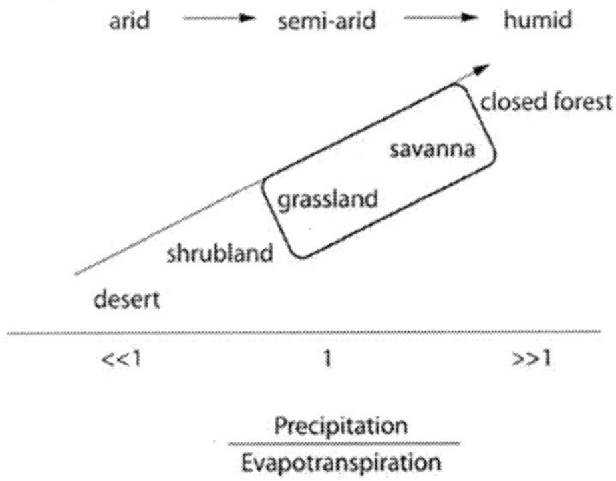

Figure: *A plot of the vegetation types that are expected to occur along a climate gradient estimated by the ratio of the incoming precipitation to the potential evaporation rate. Courtesy of Ehleringer, unpublished figure.*

Here, we talk about the two most important climate parameters influencing plant growth: water and temperature. Clearly plants will not grow in the absence of water, so the important factor influencing photosynthetic pathway distribution becomes the temperature during the growing period. Based on many systematic surveys of the natural vegetation across the globe that have been accumulating over time, a clear picture is emerging. C_4 taxa are found in warm to temperate environments and are uncommon in cool to cold climates. Below we construct a global map that describes the general abundances of C_3 and C_4 taxa on different continents.

Notice that C_4 taxa are not very common in tropical regions (0–20° latitude), because dense tropical forests tend to shade out C_4 grasses. The C_4 taxa are most common away from the tropics in the savanna and steppe regions; their abundances tends to diminish south of the desert zones (generally 30–40° latitude). Climate can be presented along an axis of an increasing of precipitation relative to evaporative water loss. At the upper, wetter end of this gradient, we would find the forests. At the drier end, where precipitation is low relative to evaporative demand, we see the deserts. In the figure below, we see that these different vegetation zones reflect variations in the ratios of precipitation-to-evaporation. C_4 plants are most common in this semi-arid region.

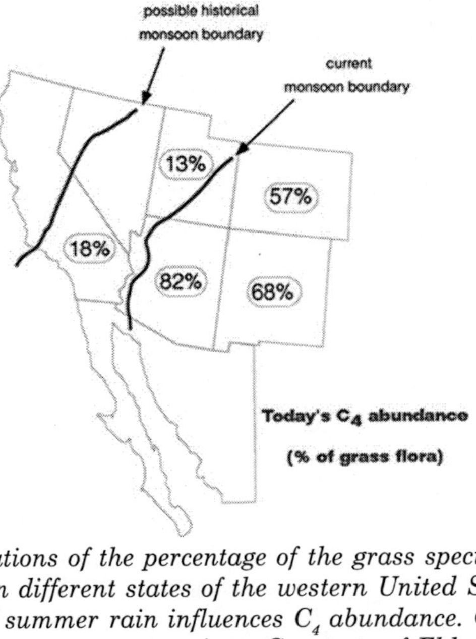

Figure: *Calculations of the percentage of the grass species that have C_4 photosynthesis in different states of the western United States. Note that the presences of summer rain influences C_4 abundance. C_4 plant tend to occur in warm, summer wet regions. Courtesty of Ehleringer, Cerling, and Dearing (2005).*

In deserts with wet monsoon summers, such as in central Africa, the sub-Indian continent, and western North America, C_4 taxa are abundant. In deserts that receive winter rainfall, such as deserts in Asia and North Africa, there are few C_4 taxa. In North America, this regional influence of summer precipitation is most visible in the southwestern United States where there are large differences in precipitation associated with the Arizona monsoon boundary as shown below.

Human activities and disturbances by animals will influence the distribution of C_3 and C_4 taxa within savannas and grasslands. This reflects the role of disturbances, such as grazing and fires, on the importance of trees on the landscape.

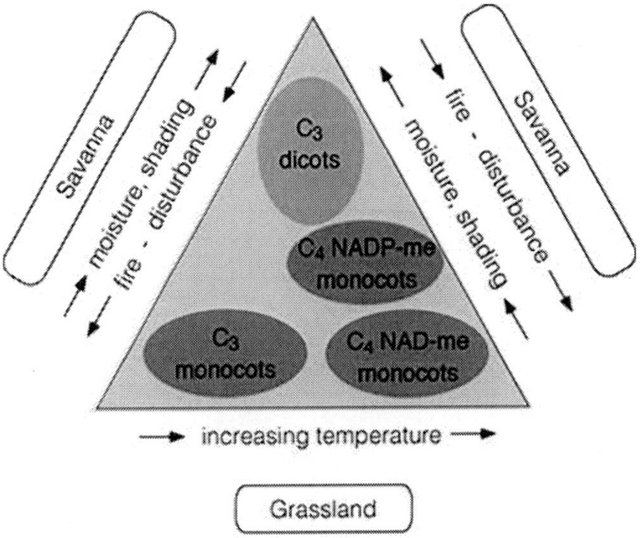

Figure: *A three-axis, triangular presentation of the different gradients influencing the abundances of different C_4 photosynthesis subtypes. Courtesy of Ehleringer, Cerling, and Dearing (2005).*

Across the grasslands and savannas, the two dominant C_4-photosynthesis subtypes do not share identical distributions. The C_4 NADP-me grasses tend to occur in drier regions, such as the shortgrass prairie of the Great Plains. On the eastern, wetter edge of the Great Plains, the shortgrass prairie is replaced by tallgrass prairie, dominated by C_4 NAD-me grasses.

Today, agricultural practices result in C_4 plants growing outside of the distributions shown above. This results from the extensive planting of corn, sugarcane, and millet in both tropical and temperate regions.

Projected Future Increases in Atmospheric CO_2

Human use of fossil fuels (coal, oil, and natural gas) continues to increase as growing human populations demand more energy for transportation, heating, and manufacturing. We measure atmospheric CO_2 in units of ppm or parts per million. The rate of atmospheric increase in CO_2 is about 3 ppm per year.

Note the cyclic nature of the atmospheric CO_2 data, in which one oscillation cycle is exactly one year. This annual pattern reflects changes in the balance of photosynthesis (decreases atmospheric CO_2) and respiration (increases atmospheric CO_2) at a location over the course of the year. Atmospheric CO_2 tends to decrease in the spring and summer, when photosynthesis rates within an ecosystem exceed respiration rates. In contrast, atmospheric CO_2 tends to increase in the fall and winter when respiration rates exceed photosynthesis. In 2007 atmospheric CO_2 reached an average value of 384 ppm and is expected to reach 400 ppm before 2015.

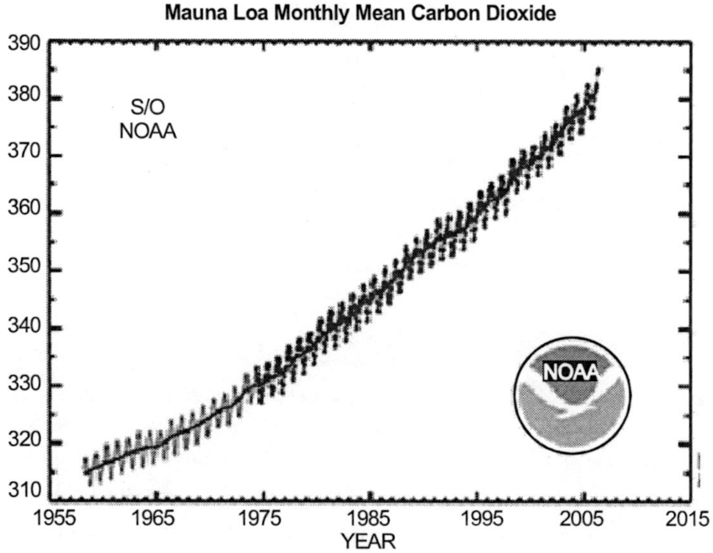

Figure: *A high precision record of the atmospheric carbon dioxide levels measured on Manua Loa, Hawaii.*

Economists have good estimates of the rate of CO_2 emission globally. The U.S., European countries, China, Japan, and India are the largest sources of fossil fuel emissions.

One surprising fact is that the observed rate of atmospheric CO_2 increase is actually less than the observed rate of atmospheric CO_2

increase. This is because plants on land and algae in the ocean are currently able to take up about one-half of fossil fuel emissions through enhanced photosynthesis. Scientists study how plants and ecosystems respond to elevated CO_2 using an experimental field approach called a Free Air CO_2 Enrichment (FACE) Experiment. In a FACE experiment, pipes inject CO_2 into the interior of a ringed area containing a complete ecosystem as shown below.

These FACE research facilities give scientists an opportunity to understand how different plant biochemical, physiological, and growth processes within the ecosystem will respond as a result of long-term exposure to elevated CO_2 levels. Since biomass production involves so much more than simply increased photosynthesis (i.e., mineral nutrients are required as well), it is doubtful that plant growth can be sustained in a linear, proportional fashion as atmospheric CO_2 levels continue to increase. The FACE studies are designed to address the question of how ecosystems will respond to future atmospheric CO_2 environments and whether the growth response level off at some future CO_2 level. Global warming and changes in climate are anticipated effects of a rapidly increasing CO_2 levels. These are, of course, but two of the many reasons why scientists and others are concerned about the consequences of elevated atmospheric CO_2. Just how much the atmospheric CO_2 will increase is unknown. Below are estimates of the ranges, based on two plausible scenarios.

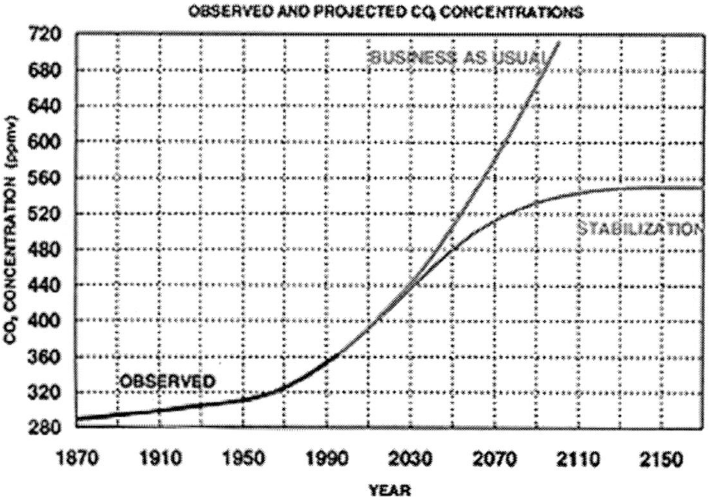

***Figure:** Projected changes in the atmospheric CO_2 concentrations under constrasting emission-control projections.*

In one scenario, titled "business as usual" atmospheric CO_2 levels are projected to reach 700 ppm by the end of this century. On the other hand, an aggressive global effort to curb CO_2 emission might result in an atmospheric CO_2 stabilization of 550 ppm.

Using Carbon Isotopes to Detect Adulteration in Foods

Carbon isotope ratios ($\delta^{13}C$) of C_4 plants such as corn and sugar cane are –10 to –12 ‰, in contrast to $\delta^{13}C$ values of C_3 foods, which are typically –26 to –28‰. The naturally occurring variations in carbon isotope ratios within the C_3 of C_4 ranges occur because environmental factors influence carbon isotope ratios. However, the differences in carbon isotope ratios among the two photosynthetic pathways are large enough to ensure that it is possible to distinguish between whether or not a food substance is based on C_3 of C_4 photosynthesis.

Adulteration is the substitution of a cheaper food substance or compound with another. Carbon isotopes are frequently used to detect the substitution of C_4 sugars for C_3 food products, such as the introduction of sugar cane into honey to increase yield. One easy way to detect this adulteration is to examine the jams and jellies that are found in restaurants. Below are the results of a survey conducted as a science project by Jeff Ehleringer when he was a young man in junior high school. Jeff collected jams and jellies every time he went into a restaurant. He had the opportunity to collect jams and jellies from across the western United States and in parts of Europe as well.

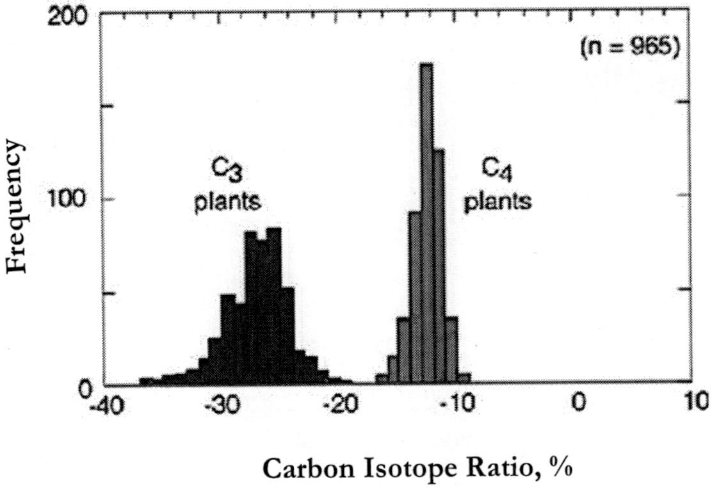

Figure: Variations in the carbon isotope ratios of C_3 and C_4 plants. Courtesy of Cerling et al. (1997) Nature.

The results shown above are interesting and indicate that adulteration has taken place in some of the samples. The fruits that go into producing jams and jellies are all C_3 fruits (grapes, strawberries, raspberries, orange, apricots, peaches, etc.) One subgroup of the jams and jellies appear to have carbon isotope ratios similar to the stable isotope ratios of C_3 plants, from which they were supposed to have been derived. These were all samples that Jeff had collected in Europe, where there are authenticity laws regulating what materials can go into making a processed food such as a jelly.

In contrast, the vast majority of the jams and jellies had carbon isotope ratios consistent with those of C_4 plants. Yet, fruits are not from C_4 plants but instead are produced from C_3 plants. These are the adulterated jellies, most likely produced largely from sugar syrup and food flavouring with very little food used in the process.

These substitutions are legal in the U.S., because here, we do not have authenticity laws with respect to food production. A third small subgroup consists of jams that were made of 50/50 C_3/C_4 food sources. It turns out that these were the more expensive jams and jellies. There are many other food products that are tested for their carbon isotope ratios to detect adulteration. Common among these are fruit juices, vanilla (vanilla beans come from CAM plants), beer (wheat and barley are C_3), and alcoholic beverages.

Reconstruction of the Expansion of C_4 Taxa

C_4 photosynthesis is favoured over C_3 photosynthesis under conditions of high temperature and/or low atmospheric CO_2. Below we provide a 3-D graphic of how the ratio of photorespiration-to-photosynthesis increases as temperature increases and/or CO_2 decreases.

The rate of photorespiration in a leaf increases because RuBP carboxylase is more likely to react with O_2 instead of CO_2 as temperature increases and/or CO_2 decreases. The result of an increase in photorespiration is a decrease in photosynthetic quantum efficiency. This leads to environmental conditions where C_4 plants are favoured over C_3 and vice versa.

This model leads to predictions that the abundances of C_3 and C_4 plants should vary spatially and as a function of atmospheric CO_2 history. The secret to evaluating this model is to go back and look at plant fossils over time and to relate photosynthesis pathway with prehistoric atmospheric CO_2. This is a difficult task.

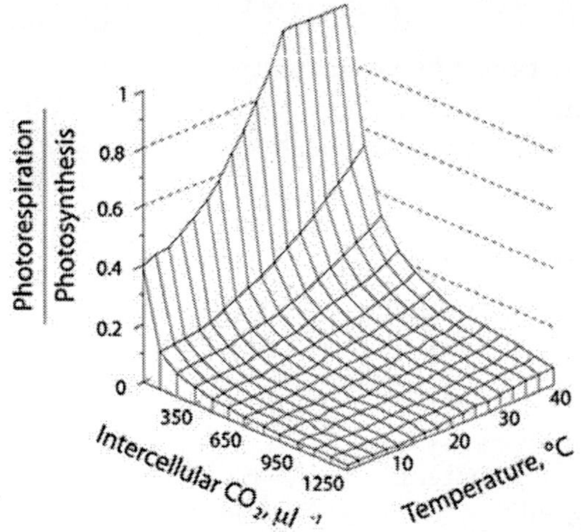

Figure: *Changes in the ratio of photorespiration-to-photosynthesis as a function of both CO_2 concentration and temperature for a plant with C_3 photosynthesis. Courtesy of Ehleringer and Cerling (2001).*

Reconstruction of the expansion of C_4 plants over the past 10–15 million years can be a challenge, because individual plants are not well recorded in the fossil record in many locations. Most plant fossils are thought to have formed as a result of submersion into an anaerobic, aquatic environment. The vast majority of plants do not become fossils under these conditions, but instead are decomposed by microbial activities leading to diminished numbers of fossils that we can use to reconstruct the expansion of C_4 plants over time. However, there are good proxies.

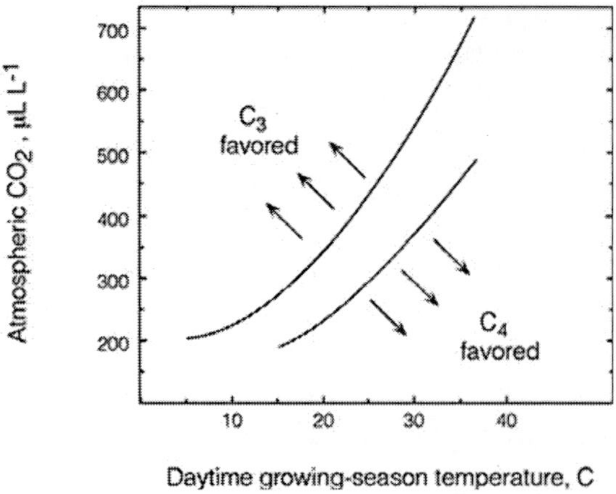

Figure: *Predicted realms of atmospheric CO_2 and growing season temperature that favour plants with C_3 or C_4 photosynthesis.*

Carbon isotope ratios ($\delta^{13}C$) of animal tissues reflect the foods that they ate. For modern animals, it is possible to choose from among, hair, muscle, bone collagen, and tooth enamel for isotope analyses in order to determine the proportions of C_3 and C_4 food sources in their diets. Hair provides a sequential record of the animal's diet during the time of hair production. Muscle and bone collagen analyses provide an integrated measure of the animal's current diet as shown in the figure below.

The δ^{13}13C of animal teeth faithfully record the carbon isotope ratios of food sources during the period of tooth development. The carbon in tooth enamel is actually carbonate that has condensed from CO_2 that was produced as a byproduct of metabolizing food. After an animal has died and decomposition has occurred, about the only remaining tissue that remains and preserves the C_3/C_4 dietary signals is the enamel in the animal's teeth.

For many herbivores, such as cows and horses, tooth growth is continuous and provides a longer-term dietary record than for animals with a fixed tooth growth period (such as humans). As teeth are preserved for millions of years, δ^{13} of tooth enamel can be used to reconstruct the abundances of C_3 and C_4 plants eaten by mammalian grazers over extended time periods. In the graphic below, we plot the frequency distributions of modern animal teeth with that observed for C_3 and C_4 plants:

Note that there is an "å" offset of 14.1‰. This is the isotope enrichment associated with CO_2 condensing to form carbonate. Now we have a tool—the carbon isotope ratio of animal tooth enamel will record the abundance of C_3 versus C_4 food sources.

Until 8 million years ago, C_4 plants were not an important part of ecosystems. Then, between 6 and 9 million years ago, C_4 plants became important parts of many ecosystems, particularly at tropical and semi-tropical latitudes. These results are consistent with model predictions above. C_4 photosynthesis is predicted not to occur on Earth until the atmospheric CO_2 level falls below a critical threshold value. Even then it should appear first in those locations with the warmest temperatures during the growing season.

There is also extensive evidence presently to show that fluctuations in the atmospheric CO_2 concentration values between glacial (180 ppm) and interglacial (280 ppm) periods were large to influence the local abundances of C_3/C_4 plants. Shown below are the time series analyses of carbon in bog in tropical Africa. At about 10,0000 to 12,000 years ago, the vegetation surrounding these bogs switched from being dominated by C_4 plants to being dominated by C_3 plants. This shift is evident by the dramatic change is carbon isotope ratios of materials feeding into the bog, especially at the time that the glacial period ended and we entered our current inter-glacial period.

Together these historical pieces of information paint an interesting history. The C_3 photosynthetic pathway dominated for much of Earth's history. And it is only relatively recently in Earth's (6–8 million years ago) history that we have seen an expansion of C_4-dominated ecosystems. What the future holds is unclear, since anthropogenic burning of fossil fuels is rapidly increasing in atmospheric CO_2 to levels far exceeding those observed on Earth over the past 1 million years.

Chapter

3

Auxin: The Growth Hormone

The Structural Requirements for Auxin Activity

Antiauxins are another class of synthetic auxin analogs. These compounds such as x-(p-chlorophenoxy)isobutyric acid, or PCIB, have little or no auxin activity but specifically inhibit the effects of auxin. When applied to plants, antiauxins may compete with IAA for specific receptors without triggering an auxin response, thus inhibiting normal auxin action. It is possible to overcome the inhibition of an antiauxin by adding excess IAA.

How can such a diverse group of chemicals all be active auxins? No other plant hormone, with the possible exception of the cytokinins, has such a wide variety of active analogs. For example, few people would have predicted that either 2,4-dichlorophenoxyacetic acid (2,4-D) or α-naphthalene acetic acid (α-NAA), both of which bear little resemblance to IAA, would be able to stimulate coleoptile elongation or have activity in other auxin bioassays.

A comparison of the compounds that possess auxin activity reveals that at neutral pH they all have a strong negative charge on the carBoxyl group of the side chain that is separated from a weaker positive charge on the ring structure by a distance of about 0.5 nm. This charge separation may be an essential structural requirement for auxin activity.

The main part of the molecule, the indole ring, is not essential for activity, although an aromatic or fused aromatic ring of a certain size range is required. Recently Edgerton and colleagues (1994) proposed a set of molecular requirements for auxin activity based on studies of the binding of various auxin analogs to a protein, auxin-binding protein 1 (ABP1), which may be the auxin receptor. Their model defines three essential regions of the binding site: a planar aromatic ring–binding platform, a carBoxylic acid–binding site, and a hydrophobic transition region that separates the two binding sites.

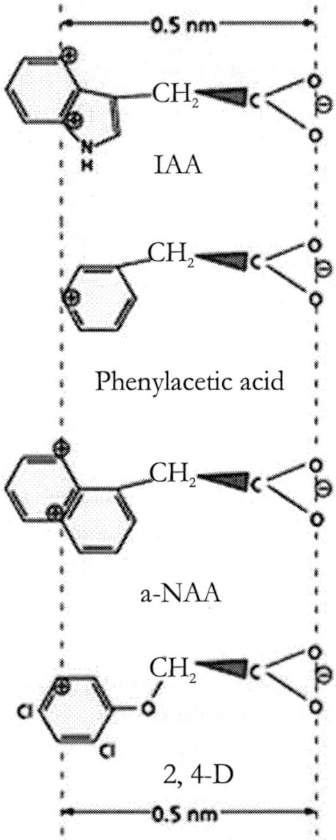

Figure: *The dissociated forms of IAA, phenylacetic acid, a-NAA, and 2,4-D (in descending order), showing the negative charge on the carboxyl group and the fractional positive charge on the ring, separated by a distance of about 0.5 nm. (After Farrimond et al. 1978.)*

Auxin Measurement by Radioimmunoassay

A radioimmunoassay (RIA) allows the measurement of physiological levels (10^{-9} g = 1 ng) of IAA in plant tissues. An RIA, requires two chemicals: (1) specific antibodies that recognize IAA; and (2) a radioactively labelled IAA in which several hydrogen atoms have been replaced by radioactive tritium. Because of the sensitivity and selectivity of the antibodies used in this technique, RIA is as sensitive as some of the best physicochemical methods. Radioimmunoassays also allow one to quantify a specific auxin. In a modified RIA, the radiolabeled IAA is replaced by an auxin conjugated to an enzyme, generally an alkaline phosphatase.

This enzyme-linked immunosorbent assay (ELISA) affords sensitivities that are similar or better than the RIA.

Evidence for the Tryptophan-Independent Biosynthesis of IAA

Further evidence was provided by labelling studies with tryptophan auxotrophic mutants of *Arabidopsis*. Together, these studies established that the branch point for IAA biosynthesis is either indole or its precursor, indole-3-glycerol phosphate. Studies with the *Arabidopsis* mutants *trp2* and *trp3*, which are blocked in the last two steps of tryptophan biosynthesis, have shown that IAN accumulates up to 11-fold in the mutants compared to the wild type. In tomato, IPA has been shown to be synthesized independently of tryptophan. Depending on the species, then, either IAN or IPA may serve as the intermediate between either indole-3-glycerol phosphate or indole, respectively, and IAA.

The discovery of the tryptophan-independent pathway has drastically altered our view of IAA biosynthesis, but the relative importance of the two pathways (tryptophan-dependent versus tryptophan-independent) is poorly understood. Some plants, such as bean seedlings, synthesize IAA primarily by the tryptophan-dependent pathway. The same is true of embryogenic carrot cell suspension cultures. However, upon removal from the medium containing 2,4-D, which initiates embryogenesis in these cultures, the cells switch over to the tryptophan-independent pathway of IAA biosynthesis. This result suggests that the specific pathway of IAA biosynthesis that is utilized may be under developmental control.

The Multiple Factors That Regulate Steady-State IAA Levels

The steady-state level of free IAA in the cytosol is tightly regulated and is determined by several interconnected processes. Auxin metabolism includes the synthesis, degradation, and conjugation of

IAA. In addition to IAA metabolism, its compartmentation in the choroplast and transport also play important roles in regulating the level of free IAA.

The amounts of IAA in different regions of a normal tobacco plant. The highest concentration of free IAA is in the apical bud. The level of free IAA in the mature regions of the stem is nearly constant. Since there is little IAA synthesis in the mature regions of the shoot, we can infer that there must be little catabolism or conjugation in these regions; otherwise the free IAA concentrations would continue to decline.

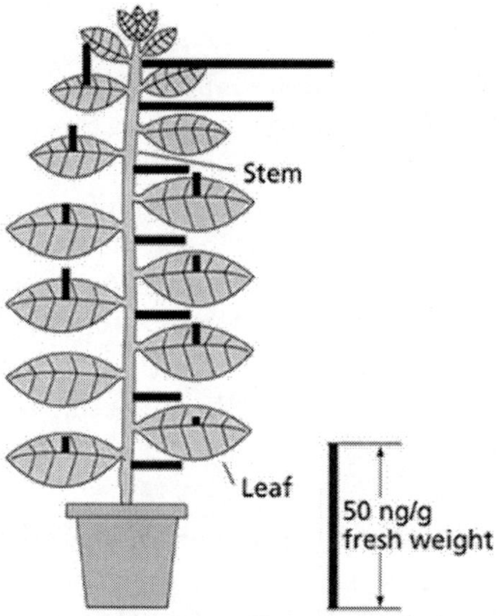

Figure: *Concentrations of IAA in different regions of the shoot of a wild-type tobacco plant. The bars at the different regions indicate the auxin concentrations found in leaves and stems.*

What happens when IAA is produced uniformly at high levels throughout the plant? To find out, tobacco plants were transformed with the two IAA biosynthesis genes from *Agrobacterium tumefaciens* so that IAA was produced throughout the plant. The level of free IAA in the upper part of the shoot increased considerably, while the free IAA concentrations in the mature parts of the shoot hardly changed. The measured levels of conjugated IAA were several times higher in the transformed tobacco plant than in the control. We can infer that processes that decrease the amount of free IAA must have been activated in the mature shoot so that the concentration of free IAA was maintained close to its normal level.

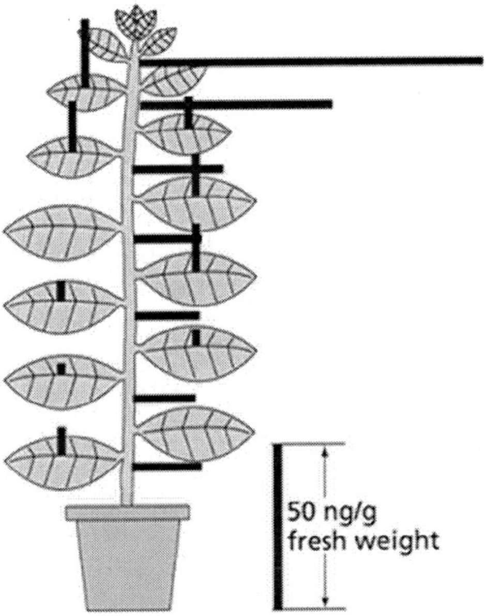

Figure: *Concentrations of IAA in different regions of the shoot of a transformed tobacco plant expressing the bacterial genes that encode the two enzymes involved in the IAM pathway of IAA biosynthesis: Trp monoxygenase and IAM hydrolase. The bars at the different regions indicate the auxin concentrations found in leaves and stems.*

The Mechanism of Fusicoccin Activation of the Plasma Membrane H^+-ATPase

Our examination of the auxin response would be incomplete without a discussion of the mechanism of fusicoccin (FC). The structure of fusicoccin A, a phytotoxin produced by the fungus *Fusicoccum amygdali*. This phytotoxin is produced by the fungus *Fusicoccum amygdali*, a parasite of peach and almond trees. Once released into the leaves, the toxin stimulates the H^+-ATPases of the guard cells. At the biochemical level it induces a rapid (10 to 30 seconds) hyperpolarization of the plasma membrane, accompanied by an acidification of the cell wall. This triggers an irreversible opening of the stomata resulting in wilting of the leaves and, eventually, the death of the tree.

FC causes membrane hyperpolarization and proton extrusion in nearly all plant tissues. Treatment of coleoptiles and stem sections with FC leads to a transient growth response, a result that provides support for the acid growth hypothesis for auxin-induced growth. But whereas auxin-induced proton extrusion has a lag time of about 10 minutes and is inhibited by cycloheximide, FC-induced proton extrusion begins after only 1 to 2 minutes, and the response is insensitive to cycloheximide. Moreover, FC-treated cells can acidify down to a much lower extracellular pH than auxin-treated cells can.

Because of these effects, FC has sometimes been referred to as a "super auxin." However, FC acts by a mechanism entirely different from that of auxin. For example, FC does not stimulate the expression of any of the auxin-induced genes, nor can it mimic the effects of IAA on other developmental processes, such as cell division. Thus the effects of FC on plants are much more limited than those of IAA, which is not surprising since FC is a toxin, not an endogenous plant hormone. In the case of proton extrusion, FC is known to activate preexisting H^+-ATPases on the plasma membrane, probably by displacing the autoinhibitory C-terminal domain of the enzyme from the catalytic site. In contrast, auxin may promote proton extrusion in part by activation of preexisting H^+-ATPases (perhaps via acidification of the cytosol) and in part by causing the de novo synthesis of the H^+-ATPase.

Fusicoccin was first identified as a phytotoxin in the 1960s, but it was not until 1977 that a radioactively labelled version of this toxin was available that could be used to characterize its ligand. Studies with radioactively labelled FC have shown that it binds tightly to a protein on the plasma membrane. Ultimately, these studies led to the purification and identification of the fusicoccin-binding protein (FCBP).

Initial attempts to purify FCBP after photoaffinity labelling led to conflicting results. Some laboratories identified a homodimer, composed of two 30 kDa subunits. The ability of the purified 30 kDa polypeptides to stimulate the plasma membrane H^+-ATPase in the presence of FC was tested in vitro proton-pumping experiments. The 30 kDa polypeptide was reconstituted along with purified plasma membrane ATPase into artificial membrane vesicles called proteoliposomes. The addition of FC to the reconstituted proteoliposomes caused an increase in their ATP-driven H^+-pumping activity.

Cloning and sequencing of the gene for the 30 kDa polypeptide revealed it to be a member of the so-called 14-3-3 family of regulatory proteins. Regulatory proteins of the 14-3-3 family are widespread among plants and animals. Originally identified in mammalian brain tissue (and named after the positions they occupy in chromatographs), 14-3-3 proteins make up a heterogeneous family of soluble proteins ranging from 25 to 32 kDa, which associate to form dimers. The precise function of 14-3-3 proteins has not yet been elucidated. However, the numerous functions that have been attributed to them indicate that they play multiple roles in signal transduction pathways.

Although it was initially believed that the 30 kDa 14-3-3 protein represented the complete FCBP, another laboratory reported that a 90 kDa polypeptide consistently copurified with the two 30 kDa polypeptides. More recently it has been demonstrated that the 14-3-3 protein is unable to bind FC by itself. To bind FC, the 14-3-3 protein requires the presence of H^+-ATPase. This finding confirms the earlier report that a 90 kDa polypeptide (the H^+-ATPase) copurifies with the 14-3-3 protein. According to the current model, FC activates the H^+-ATPase by stabilizing a transient complex that forms between the autoinhibitory C-terminal domain of the H^+-ATPase and the 14-3-3 dimer locking the enzyme into its most active state.

The Fluence Response of Phototropism

Phototropism involves two fundamental processes: blue-light perception and differential cell elongation. The identity of the blue-light photoreceptor involved in phototropism. There is now compelling evidence that two flavoproteins associated with the plasma membrane (PHOT1 and PHOT2) are the photoreceptors for phototropism. These proteins become phosphorylated in response to blue light. The differential cell elongation in response to unilateral light is thought to be caused by the lateral redistribution of auxin to the shaded side.

If phototropism were a simple photochemical reaction, one would expect the fluence response curve to rise from a threshold level of light to a plateau at which the photoreceptor becomes saturated. However, this is not what happens. The figure shows a diagram of a typical fluence response curve, which is similar to what has been observed in a wide range of species, including *Avena* coleoptiles and *Arabidopsis* stems.

Starting at the light threshold, there is a bell-shaped curve that has been termed the first positive curvature. The first positive curvature is followed by a neutral zone in which little or no measurable response is observed. Under certain conditions, *Avena* coleoptiles may actually bend away from the light in this range, a phenomenon referred to as the first negative curvature. Both the first positive and the first negative curvatures (neutral zone) are restricted to the coleoptile tip. At higher light fluences the curve ascends again, and this second bending response is termed the second positive curvature. The second positive curvature differs from the first positive curvature in that the bending occurs at the base rather than at the tip of the coleoptile.

Plant physiologists have long puzzled over the meaning of this complex relationship between light fluence and phototropism. Most studies of phototropism have been carried out in the range of the first positive curvature. The first positive curvature and the neutral zone obey the Bunsen–Roscoe reciprocity law, which states that the response will be the same for a brief exposure to bright light as for a longer exposure to dim light, as long as the product of time and the photon fluence rate are the same. Thus the first positive curvature is a response to short light pulses.

In contrast to the first positive curvature, the second positive curvature is a response to irradiations of long duration. The second positive curvature does not obey the reciprocity law; instead, it is proportional to the duration and the photon fluence rate. For the second positive response to occur, both the time threshold and the fluence threshold must be exceeded. Since these are the conditions that prevail normally in nature, phototropism in nature usually represents the second positive curvature.

What causes the gradient in auxin to form at the three fluences? It has long been assumed that the light gradient that is established across the coleoptile during unilateral irradiation with blue light must generate a biochemical gradient between the cells of the shaded and the irradiated sides, but the nature of this biochemical gradient has, until recently, eluded detection. With the discovery of blue light–induced phosphorylation of the PHOT1 and PHOT2 plasma membrane proteins, much work has focused on these photoreceptors.

Recently, Michael Salomon and his colleagues at the University of Munich demonstrated that unilateral blue light causes a lateral gradient in the phosphorylation of PHOT1 from the irradiated to the shaded side. Moreover, the concentration of PHOT1 declines exponentially from the tip of the coleoptile to the base. Studies on the rates of phosphorylation of the 116 kDa protein PHOT1 at the tip versus at the base of oat coleoptiles have led to a new model for the fluence response curve of phototropism.

According to this new model proposed by Salomon and colleagues, the gradient in concentration of NPH1 from tip to base makes the tip more sensitive to light than the base is. Thus, at threshold fluences, a gradient in phosphorylation occurs only at the tip, leading to a small bending response. Increasing the fluence increases the gradient of phosphorylation at the tip, until phosphorylation on the irradiated side saturates and a maximum phosphorylation gradient is obtained. This maximum gradient represents the peak of the first positive curvature.

As the fluence is increased beyond this point, phosphorylation increases on the shaded side due to the penetration of light, reducing the gradient. When phosphorylation of NPH1 is equal on both sides of the tip, no bending occurs (the neutral zone).

However, with continued exposure to light of sufficient fluence, phosphorylation increases at the base on the irradiated side. Phosphorylation at the base is dependent on both fluence and time, unlike phosphorylation at the tip which is strictly dependent on fluence. Thus, phosphorylation at the base appears to involve some type of time-dependent biochemical process. Since recent immunological studies have shown that the total amount of the 116 kD protein remains constant during phototropism, the time-dependent increase in phosphorylation at the base does not involve synthesis of new PHOT1 protein. Instead, a second photoreceptor (PHOT2), initially implicated in chloroplast movement in response to light, appears to function as an additional photoreceptor in second positive curvature conditions.

The model proposed by Salomon and colleagues requires further testing, and other interpretations are certainly possible. Nevertheless, the new model provides the first plausible explanation for the fluence response curve of phototropism. However, even if this model is confirmed, the question remains as to how a gradient in phosphorylation can bring about a gradient in auxin.

Gravity Perception without Statoliths in *Chara*

An alternative mechanism of gravity perception that does not involve statoliths has been documented in the giant-celled fresh water alga *Chara*. Despite the absence of any discernible sedimenting statoliths, *Chara* filaments show negative gravitropism. To explain this phenomenon, it has been suggested that the entire protoplast acts like a statolith, and that gravity is perceived at the interface between plasma membrane and cell wall.

According to the Plasmalemma Central Control (PCC) model, stretch-activated calcium channels are clustered around attachment centres connecting the cytoskeleton and the cell wall. These calcium channels are thought to open in response to tensions at the membrane caused by gravity-induced changes in the distribution of forces exerted by the protoplast, the cytoskeleton, or the cell wall. Something like the PCC model may explain how *Chara* and the starchless mutants of *Arabidopsis* are able to perceive and respond to gravity.

Cytokinins: Regulators of Cell Division

Cultured Cells Can Acquire the Ability to Synthesize Cytokinins

When cultured normal callus tissues of many species are subcultured repeatedly over a long period, they may become hormone autonomous; that is, they will grow on culture medium lacking an auxin or a cytokinin. This phenomenon is known as habituation.

There are several possible explanations for the habituation of plant tissues.

First, the tissues may acquire the capacity to synthesize their own cytokinin, and in some cases habituated callus tissues have been shown to contain cytokinins. Alternative explanations are that the rate of cytokinin destruction is reduced, or that the tissues become more sensitive to cytokinins, such that lower levels of the hormone are effective.

Frederick Meins and his colleagues in Basel, Switzerland, have shown that different tissues of tobacco plants exhibit different capacities to synthesize cytokinins in tissue culture. Tobacco pith tissue is cytokinin dependent but capable of being habituated. In contrast, tissues cultured from the cortex of the stem are cytokinin autonomous; they can be cultured indefinitely on medium containing auxin as the only hormone supplement. Finally, tobacco leaf tissue has an absolute requirement for cytokinins and cannot be habituated.

These observations indicate that the ability to synthesize cytokinins can be switched on or off during development, and that the on and off states can be stably maintained during cell divisions. Such developmental control of cytokinin synthesis may be important in morphogenesis.

Various Methods Are Used to Detect and Identify Cytokinins

Naturally occurring molecules with cytokinin activity may be detected and identified by a combination of physical methods and bioassays. The introduction of immunoaffinity purification methods combined with liquid chromatography and mass spectrometry, along with the use of isotopically labelled cytokinins as internal standards, has resulted in major advances in the measurement of endogenous cytokinins.

Historically, bioassay was the only method available for the identification of molecules that act as cytokinins. Cell proliferation bioassays using tobacco pith tissue or carrot taproot continue to be important because the ability of these tissues to initiate and sustain cell proliferation is proportional to the concentration of active cytokinin in an extract, provided the extract does not contain growth inhibitors.

In addition, some continuously cultured callus tissues of tobacco and soybean that cannot grow in culture without cytokinin have been used for cytokinin bioassays. All of these cytokinin-requiring tissues exhibit a linear increase in growth with increasing cytokinin concentration over a fairly broad range, although high concentrations of cytokinin usually inhibit growth.

Immunological methods are very useful for cytokinin identification and quantification. Researchers can produce antibodies against cytokinins by injecting rabbits or mice with cytokinin ribosides conjugated to a protein. Monoclonal antibodies also have been generated that are highly specific for individual cytokinins. These antibodies can be used to quantitate the amount of a cytokinin in a sample by means of a radioimmunoassay.

For cytokinin isolation, plant extracts are first fractionated, usually by high-performance liquid chromatography (HPLC), and the cytokinins in the fractions are detected and measured by means of a cytokinin radioimmunoassay, similar to the auxin radioimmunoassay.

The cytokinin antibodies can also be used to isolate the hormone from extracts by immunoaffinity chromatography. Immunological methods hold great promise for the identification and quantification of naturally occurring cytokinins because the antibodies are highly specific and more sensitive than most bioassays. Furthermore, these immunological methods are very rapid.

Cytokinins Are Also Present in Some tRNAs in Animal and Plant Cells

Most organisms contain a variety of modified bases in their tRNA. The normal RNA bases are incorporated into the newly synthesized tRNA chain, and then specific enzyme modify a small subset of the bases. As with the free cytokinins, Δ^2-isopentenyl groups are transferred to the adenine molecules. The enzyme catalyzing this reaction is called tRNA-IPT, which is weakly similar in its sequence to the plant cytokinin biosynthetic enzyme. The similarity was used to help identify the cytokinin biosynthetic enzyme from plants. The tRNA-IPT must be able to recognize a specific base sequence in the tRNA and transfer the isopentenyl group to the adenosine nearest the 3' end of the anticodon. It does not utilize free AMP as a substrate.

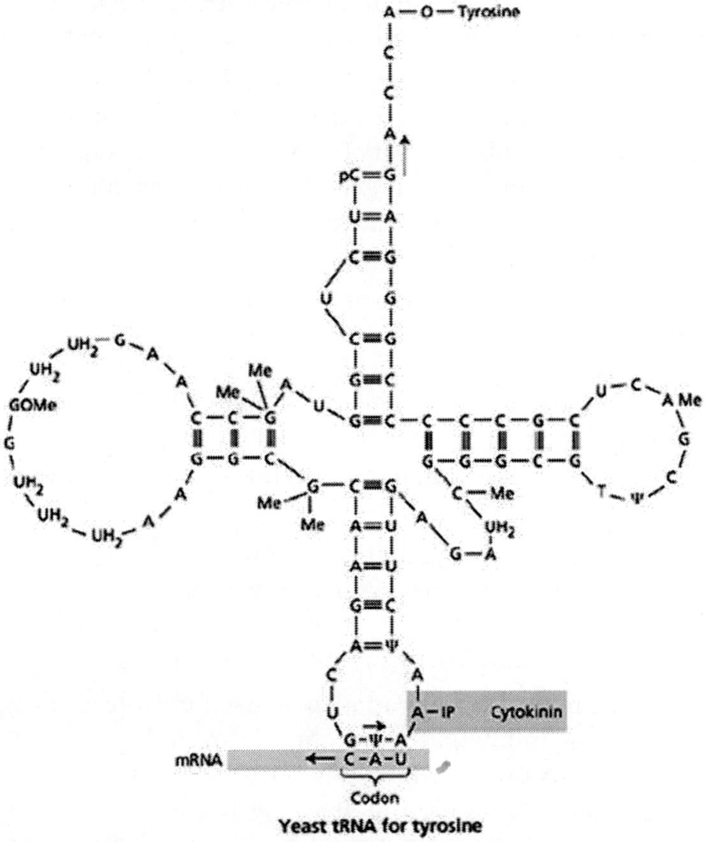

Yeast tRNA for tyrosine

Figure: *Some tRNAs contain a cytokinin. The cytokinin in the examples shown is]^{9}R] iP. Although cytokinins are present in tRNA, tRNA is probably not the source of free cytokinins in plants.*

The possibility that the free cytokinins are derived from tRNA has been explored extensively. Although the tRNA-bound cytokinins can act as hormonal signals for plant cells if the tRNA is degraded and fed back to the cells, it is unlikely that any significant amount of the free hormonal cytokinin in plants is derived from the turnover of tRNA.

The Ti Plasmid and Plant Genetic Engineering

The Ti plasmid (tumour-inducing plasmid) of *Agrobacterium tumefaciens* has been developed as a vehicle for introducing foreign genes into plants. When *Agrobacterium* infects plants, a region of the Ti plasmid called the T-DNA is taken up by the plant cell and incorporated into one of its chromosomes.

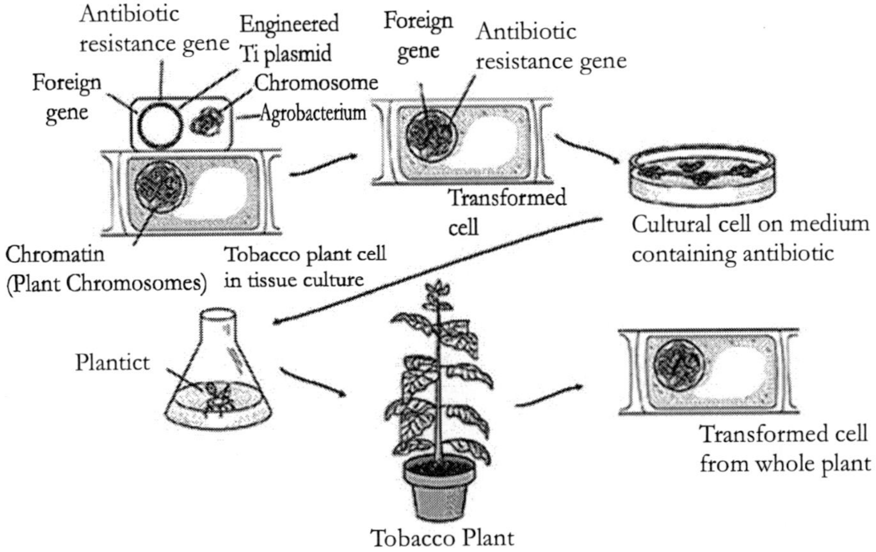

The genes in the T-DNA are referred to as phyto-oncogenes because they induce neoplastic, or tumour-producing, growth (Matzke and Chilton 1981). To use the Ti plasmid as a vector for introducing new genes into plants, it is necessary to disarm the plasmid so that it does not cause tumours. Researchers accomplished this task by deleting the genes in the T-DNA that encode the enzymes controlling auxin and cytokinin synthesis. In addition, it is necessary to introduce a gene into the T-DNA that will enable the investigator to select the transformed cells. Genes for antibiotic resistance are normally used for this purpose.

A cloned gene can then be inserted into the T-DNA of the engineered Ti plasmid, and the plasmid can be used to infect cultured cells, leaf discs, or root slices. The infected cells are placed on a culture medium that contains auxin and cytokinin (to induce growth) and the antibiotic. Only the transformed cells can grow in the presence of the antibiotic, because they have received the T-DNA containing not only the foreign gene but also the gene for antibiotic resistance.

To obtain a plant containing the foreign gene, it is necessary to regenerate whole plants from the cultured, transformed cells. Fortunately, methods for accomplishing this regeneration have been developed for many plants, although not for all important crop species yet. The regeneration involves adjusting the ratio of cytokinin to auxin to stimulate both shoot and root formation.

Although the transformation of cereal crops with *Agrobacterium* and their regeneration from cultured cells has been difficult, some remarkable successes have resulted from this approach. Investigators have introduced numerous foreign genes into plants such as tobacco, including soybean storage protein genes. Genes for such desirable characteristics as disease resistance, herbicide resistance, and salt tolerance have been transferred to crop plants by these techniques, and commercial crops now are being grown with these genetically engineered strains.

Arabidopsis can also be transformed directly by exposing whole plants to a solution containing *Agrobacterium* that is carrying engineered or wild-type Ti plasmids. The plants must be treated in such a way to allow the *Agrobacterium* to enter tissue, either by applying a vacuum or by treating with detergents. The *Agrobacterium* penetrates the floral tissue and transforms the developing ovules. Isolation of seeds from these *Agrobacterium*-exposed plants yields up to 2% of the seeds that are transformed with the T-DNA. This approach is very useful for molecular genetic studies, such as for characterizing DNA sequences involved in the control of gene expression, or constructing large libraries of insertional mutants.

Cytokinin Can Promote Light-Mediated Development

The *det* mutants of *Arabidopsis* develop many of the characteristics of light-grown seedlings when they are germinated in the dark. The *det* mutants have short, unhooked hypocotyls and expanded cotyledons, and the apical meristem initiates leaves that subsequently expand without light. Although the *det* mutants do not develop full photosynthetic competence in the dark, there is partial development of the chloroplasts, including the expression of genes encoding many photosynthetic enzymes that are not present in the chloroplasts of dark-grown wild-type seedlings.

When wild-type *Arabidopsis* seedlings are germinated in darkness in the presence of cytokinin, they develop many of the characteristics of the *det* mutants: Hypocotyls shorten, cotyledons expand, the apical meristem initiates leaves, and there is partial development of the chloroplasts, including the synthesis of some photosynthetic enzymes. This effect is proportional to the cytokinin dosage given to the seedlings. Cytokinin is said to produce a *phenocopy* of the *det* mutations.

The ability of exogenous cytokinin to cause de-etiolation of dark-grown seedlings is mimicked by certain mutations that lead to cytokinin overproduction. Dark-grown *Arabidopsis* seedlings carrying the *amp1* mutation have zeatin levels that are approximately 2.5 times higher than those of the wild type and develop the characteristics of *det* mutant seedlings. Although the *det* mutants do not have elevated levels of cytokinins, they may have altered sensitivity to cytokinins, since *Arabidopsis* tissues from *det1* plants, when cultured in vitro, do not require cytokinin for growth and become green.

Although the functions of the genes altered by the *det* mutants are not completely understood, the *det2* mutation is in a gene required for the synthesis of brassinosteroids. Brassinosteroids and cytokinins may have antagonistic effects in regulating light-induced development. These results suggest that cytokinins may be part of the light signal transduction pathway that is responsible for the initiation of normal vegetative development and photosynthetic competence. However, the role for cytokinins in the developmental pathway initiated by light is not clearly established.

Localization of Cytokinin Signalling Elements

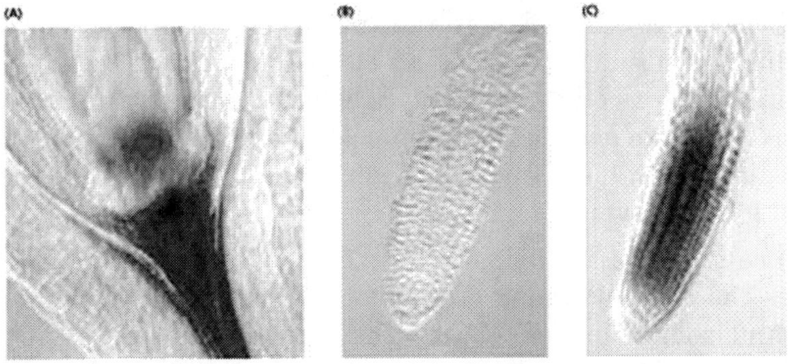

Figure: *Expression of* ARR5. *The pattern of* ARR5 *expression was examined by fusion of the promoter to a* GUS *reporter gene (A) or by whole-mount* in situ *hybridization (B and C). For the latter, the tissue is hybridized with labelled single-stranded* ARR5 *RNA in either the sense orientation (B) or the antisense (C). The sense RNA is a negative control and reveals background, nonspecific staining. The antisense probe specifically hybridizes with the* ARR5 *mRNA present in the tissue, thereby revealing its spatial distribution. With both methods,* ARR5 *expression is observed primarily in the apical meristems.*

Expression of *ARR5*. The pattern of *ARR5* expression was examined by fusion of the promoter to a *GUS* reporter gene (A) or by whole-mount *in situ* hybridization (B and C). For the latter, the tissue is hybridized with labelled single-stranded *ARR5* RNA in either the sense orientation (B) or the antisense (C). The sense RNA is a negative control and reveals background, nonspecific staining. The antisense probe specifically hybridizes with the *ARR5* mRNA present in the tissue, thereby revealing its spatial distribution. With both methods, *ARR5* expression is observed primarily in the apical meristems. (From D2 Agostino et al. 2000.)

Ethylene: The Gaseous Hormone

Cloning of the Gene That Encodes ACC Synthase

A major hurdle in the understanding of ethylene biosynthesis was the difficulty in purifying the low abundance, unstable protein and the cloning of the corresponding gene. Researchers overcame this obstacle by raising antisera against partly purified ACC synthase from zucchini fruit slices that were induced to produce large amounts of ethylene. The resulting antisera contained a mixture of antibodies, including antibodies to ACC.

The investigators then removed the contaminating antibodies by allowing them to bind to proteins extracted from zucchini tissue that lacked the enzyme. The key to the success of this procedure was the availability of tissue largely lacking the enzyme with which to subtract the non-ACS antibodies from the polyclonal antibody mixture. This circumvented the need for highly purified ACC synthase protein. The purified ACC synthase antibodies were then used to isolate the gene encoding the enzyme.

Once the gene encoding ACS was isolated, it could be used to express the ACC synthase protein in the bacterium *E. coli* at high levels and isolate purified ACC synthase in large amounts. The cloned zucchini gene also allowed genes encoding ACC synthase to be isolated from numerous other plant species, using low stringency hybridization and degenerate PCR.

The deduced amino acid sequences of ACC synthase genes were found to show a high degree of similarity to the *aminotransferase* superfamily of enzymes. These enzymes catalyze the transfer of amino groups among amino acids and require pyridoxal phosphate as a cofactor.

Demonstration of the Dimeric Nature of ACC Synthase

The deduced crystal structure of both apple and tomato ACS reveals a homodimeric protein with shared active sites, which is similar to aminotransferases. The dimeric nature of ACC synthase has been confirmed by an elegant study involving functional complementation. In this study, researchers examined potential dimerization of ACS enzymes by expressing in *E. coli* two ACS proteins that harbour distinct active-site mutations, neither of which has activity on its own. If ACS acts as a monomer, then expressing the two different mutant versions should not result in an active enzyme, because each is inactive. Conversely, if the two inactive ACS subunits come together as a dimer, the different amino acid changes can complement each other, forming a functional enzyme. The researchers indeed found that functional ACS enzyme was formed when various pairs of mutants were expressed, which provides compelling evidence that ACS enzymes are able to act as both homo-and heterodimers with shared active sites.

Cloning of the Gene That Encodes ACC Oxidase

The gene that encodes ACC oxidase was cloned by a circuitous route. Researchers isolated a set of genes that are highly induced during tomato fruit ripening. They then blocked the expression of these genes by introducing an antisense version of each into transgenic tomatoes. Blocking of one such gene, *pTOM13*, repressed fruit ripening and inhibited ethylene production. Expression of *pTOM13* in *Xenopus* oocytes as well as in yeast conferred the ability to convert ACC to ethylene, indicating that the gene encodes an ACC oxidase enzyme. Using this gene, researchers then isolated ACC oxidase clones from numerous plant species.

ACC Synthase Gene Expression and Biotechnology

As we have seen, ACC synthase is a cytosolic enzyme that catalyzes the first committed step in ethylene biosynthesis in higher plants. It is a key regulatory enzyme that catalyzes the production of the ethylene precursor ACC from AdoMet (*S*-adenosylmethionine). However, this enzyme is difficult to purify because it is labile and present in low abundance in plant tissues. The molecular cloning and functional expression of ACC synthase in *E. coli* and yeasts have facilitated biochemical and structural studies of this enzyme.

ACC synthase cDNAs and genomic sequences have been cloned from numerous plant species. The emerging picture from the study of these genes is that ACC synthase is encoded by members of a divergent multigene family in which each gene is differentially regulated by various environmental and developmental factors during the growth of a plant. *Arabidopsis*, for example, has twelve and tomato has at least nine gene members, which are differentially regulated by inducers such as auxin, fruit ripening, and wounding.

The protein synthesis inhibitor cycloheximide is a potent inducer of many ACC synthase genes. Cycloheximide inducibility is a characteristic of *primary response genes* such as the early auxin-inducible genes. Two mechanisms have been proposed to explain this inducibility.

- First, the transcription of these genes may be under the control of a shortlived repressor protein whose synthesis is prevented by the addition of cycloheximide;
- Second, the mRNA transcripts of these genes may be shortlived and are stabilized when cycloheximide prevents the synthesis of a nuclease.

The induction of at least one of the genes for ACC synthase (*ACS2*) in *Arabidopsis* by cycloheximide appears to be due to transcriptional activation, and thus is likely regulated by a shortlived transcriptional repressor.

A clue to the signal transduction mechanism involved in the regulation of ACC synthase gene expression has come from experiments with lithium (Li^+). Li^+ is known to interfere with the phosphoinositide signalling pathway by inhibiting the activity of inositolphosphate phosphatase. Li^+ is also one of the strongest inducers of ACC synthase activity in plants. This finding suggests that the regulation of at least some ACC synthase genes may involve Ca^{2+} mobilization mediated by inositol trisphosphate (IP_3).

Because ethylene plays such an important role in fruit ripening, the cloning of the genes for ACC synthase and ACC oxidase has enabled scientists to use biotechnology to reduce ethylene production during fruit ripening, and thus to prolong the storage life of fruit. In the case of tomato, two different strategies have been employed to control ethylene production.

- The first strategy was to use the Ti plasmid of *Agrobacterium tumefaciens* to transform tomato tissue with antisense DNA for ACC synthase or ACC oxidase. When the genes for the two enzymes are expressed in the reverse direction in the cell, the resulting "antisense transcript" binds to and inactivates the "sense transcript" produced by the wild-type gene, effectively shutting down expression of the wild-type gene. After transformation, intact plants can be regenerated in tissue culture.

- In the second strategy for reducing ethylene production, tomato tissue was transformed (using *A. tumefaciens*) with genes encoding enzymes that metabolize ACC or AdoMet. In this case, the ACC that is produced is rapidly degraded, and no ethylene can be synthesized.

Regardless of the strategy used, all fruits from these genetically engineered plants show significant delays in ripening. Tomato fruits from plants transformed with the ACC synthase antisense DNA fail to ripen unless treated with ethylene.

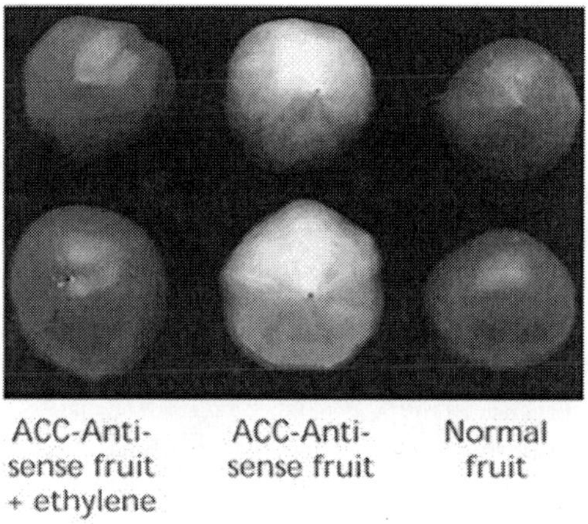

Figure: *Inhibition of tomato fruit ripening in detached tomato fruits by antisense ACC synthase RNA. Antisense tomato fruits (centre) remain firm and green unless treated by ethylene. Ethylene-treated antisense fruits (left) are indistinguishable from naturally ripened fruits (right) in terms of colour, flavour, aroma, and texture. (Reprinted with permission from Oeller et al. 1991.)*

Similar approaches are now being carried out in other agronomically important crops, such as melon and banana. In addition, these approaches are being used to retard floral senescence in carnations, thereby reducing postharvest losses. The research on ACC synthase and other ethylene biosynthesis genes thus provides an excellent example of how basic research can lead to improvements in agricultural and horticultural crops, the full benefits of which we are only beginning to realize.

Abscission and the Dawn of Agriculture

Civilization as we know it is entirely dependent on an abundant food supply, which can be provided only by agriculture. Before the development of agriculture, people lived in small, nomadic groups whose migration patterns followed the seasonal availability of foods derived from wild plants and animals. The shift from hunting and gathering to cultivation and pastoralism led to the establishment of permanent settlements with ever-expanding populations. This transition, considered to be the most significant (and by some, the most disastrous!) cultural change in human history, was made independently at least seven times at various locations around the world.

Remains of some of the earliest settlements are found in the region of the Near East called the Fertile Crescent, a broad arching zone of grassland and woodland that begins at the eastern edge of the Mediterranean and curves around some 2000 km eastward to the Zagros Mountains. One of the oldest known agricultural settlements in the world is found here, in the ancient city of Jericho, located 14 miles north of the modern city of Jericho in the Jordan Valley. Here archaeologists have unearthed evidence for a transition from hunting and gathering to agriculture that occurred around 10,000 years ago.

The first step in the transition from hunting and gathering to agriculture was the domestication of plants and animals. The seven primary domesticates of the Fertile Crescent were barley, emmer wheat, einkorn wheat, sheep, goats, cattle, and pigs. The first of these to be domesticated were the cereals. The Mediterranean region is particularly well stocked with cereals suitable for domestication. Of the world's 56 species of large-seeded grasses, 32 grow wild in the Mediterranean area. Recent DNA analyses have pinpointed the area near the Karacadâg mountains in Turkey as the probable site of origin of einkorn wheat.

All the evidence suggests that wheat and barley were domesticated over a period of only 2 or 3 centuries, between 10,000 and 9700 B.P.. Considering the large number of morphological and physiological differences between even the most primitive of the domesticated forms of wheat and barley today and their wild progenitors, it is difficult to envision how so many changes would come about in such a brief period by natural selection alone. Clearly, some type of artificial selection had to be carried out (most likely by women) to accelerate the evolutionary process. Archaeologists and botanists have attempted to reconstruct this selection process.

Of all the anatomical and physiological differences between wild and domesticated cereals, two are regarded as most crucial. The first is the loss of seed dormancy—the failure of certain seeds to germinate when planted in moist soil unless given additional environmental signals, such as cold or light. The process of plant domestication probably began in small garden plots planted near temporary settlements. It is easy to imagine how the loss of seed dormancy might be selected for by a nomadic group, since seeds that germinated and grew rapidly would be more likely to have been gathered than those whose germination was delayed.

The second critical change that took place during cereal domestication was the development of a tough rachis. The rachis is the main axis of the inflorescence, or spike, of wheat and other cereals, to which the spikelets are attached. Spikelets are small inflorescences bearing one or more florets, or small flowers, along with a set of miniature bractlike leaves. After pollination, the fruits (caryopses) develop within the spikelets. When the caryopses of the wild-type wheat are fully ripened, a series of abscission layers forms that divides the rachis into dispersal units consisting of a single spikelet attached to a short segment of the rachis. This disarticulation of the rachis begins at the top of the inflorescence and works its way downward. A ripe head of wild wheat is thus easily shattered into dispersal units when touched or blown by the wind and is referred to by archaeologists as a *brittle rachis*.

A brittle rachis is an extremely undesirable trait from a human standpoint. Imagine the plight of a small nomadic group that carefully timed its peregrinations to coincide with the ripening of wild cereals. Because of the extreme fragility of the rachis, it would be crucial to arrive close to the time of ripening, but before disarticulation of the rachis. A slight miscalculation could spell disaster for the group. Arriving too late, they would be greeted by a field of barren stalks yielding only a handful of remnant grains. As a consolation, however, they might find the rare mutant spikes that remained intact because the abscission layers of the rachis failed to develop. This property is referred to by archaeologists as a *tough rachis*. If mutants with the tough-rachis phenotype were then gathered and planted in gardens, their numbers would, over time, increase.

A second scenario for the selection of a tough rachis involves the method of harvesting. The most efficient method for harvesting wheat is to cut it off at the base of the stalk, and evidence suggests that flint-bladed sickles were used for this purpose. However, harvesting wild wheat by cutting would have to be performed before formation of the abscission layers, or the fragile spikes would shatter, sending their spikelets flying to the ground. This method of harvesting would therefore favour the selection of mutations that delayed the abscission process. Hillman and Davies (1990a, 1990b) have demonstrated in field experiments that over many generations, the population would come to be dominated by the varieties with tough rachises. Once cut, the stalks of these varieties of wheat would then be transported to another location, where the grains would be removed by threshing.

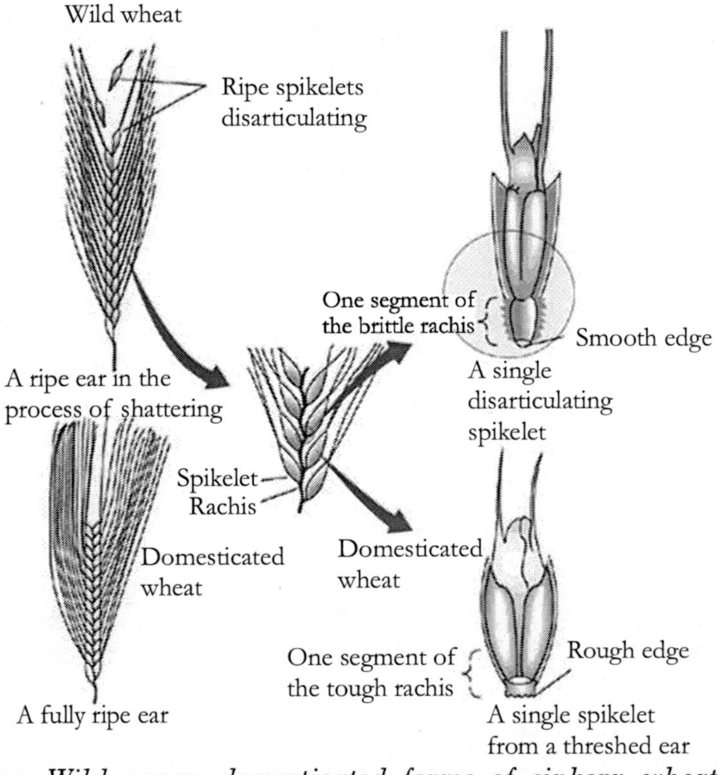

Figure: *Wild versus domesticated forms of einkorn wheat. The spikelets of the wild variety abscise readily from the ripened head at abscission zones located nearby on the rachis, leaving a smooth abscission scar on the disarticulated spikelet. In contrast, the abscission zones of the domesticated varieties that have tough rachises fail to develop properly, and the spikelets remain on the head until released by threshing. Note that the breakage of the rachis below the spikelet leaves a rough edge, which tends to impede penetration into the ground.*

Is the mutation rate in wild cereals rapid enough to account for their domestication in a period of only 200 to 300 years? Sharma and Waines (1980) have determined that two genetic loci are involved in the evolution of the tough-rachis character in einkorn wheat. Even if we assume that a mutation that converts a brittle rachis to a tough rachis occurs in only one out of every million wild-type wheat plants, Hillman and Davies (1990a, 1990b) have estimated that varieties with tough rachises would become the dominant forms in wild wheat populations after only 200 years of human management. Higher rates of mutation would shorten this time considerably, possibly to within 20 years.

Thus the experimental results and calculations are in good agreement with the archaeological evidence. Agriculture in the Fertile Crescent, which ultimately led to the founding of Greece and Rome and all of Western civilization, began with mutations in genes that control abscission.

Conservation of Ethylene Signalling Components in Other Plant Species

Genes that are similar to the *Arabidopsis* ethylene signalling components have been found in other plant species. For example, the tomato genome, like that of *Arabidopsis*, encodes a family of ethylene receptor genes, including *Never-ripe*, which are differentially expressed throughout development. The tomato ethylene receptor genes are homologous to ETR1. A tobacco homolog of EIN3 has been found, and genes similar to all the *Arabidopsis* signalling components can be found in the rice genomic sequence. These and other studies suggest that the components of these proteins probably represent ubiquitous mediators of ethylene signalling in plants.

Chapter

4

Respiration and Lipid Metabolism

Although it is possible to measure respiration by intact tissues it is often useful to be able to isolate uncontaminated, intact and functional mitochondria. Here, we will consider some of the principles for isolating cell organelles. This discussion is not meant to be exhaustive nor will it be a cookbook with recipes to be followed step-by-step.

Choice of Material

If the aim of the experiment is to investigate a specific physiological problem, then the plant material is often well-defined. However, in many cases there is some flexibility in the choice of material and it is therefore wise to follow certain rules of thumb:

- Choose a tissue that is easy to obtain in relatively large amounts (at least 100–200 g fresh weight), such as tubers or hypocotyls.
- Choose a tissue for which a good protocol is available. It can take quite a lot of time and effort to optimize an isolation protocol.

Note that the isolation of mitochondria from green tissues presents special problems because of the large amount of chloroplasts. Roots are a notoriously difficult tissue for isolation of mitochondria.

Disruption of the Tissue

The aim is to rupture as many of the cells in the tissue as possible in order to release their cytoplasm with the mitochondria. Whereas many mammalian tissues are easy to homogenize, plant tissues are much tougher mainly because of the cell walls. There are a number of standard homogenizers, including mixers with rotating knives or razor blades, juice extractors, ball mills, and mortar and pestle; the most suitable method for the specific plant material should be chosen. This may involve some initial optimization experiments.

Homogenization Medium

The medium for homogenizing rat liver only contains an osmoticum and a buffer. That will not work with plant tissues because of the disruption of the large central vacuole (absent in mammalian cells), which releases its fairly acidic content often containing phenols and other potentially harmful substances.

Let us consider the components of the optimal homogenization medium for isolating intact and functional organelles from plant material, one by one.

Osmoticum. This is required to prevent the mitochondria from losing their matrix content because of swelling and bursting. For the isolation of plant mitochondria, there is a tradition of using sucrose or mannitol at 0.2–0.5 M; whereas 0.15 M KCl is standard for the isolation of mammalian mitochondria. In neither case does this mimic precisely the conditions in the cytosol, but it works.

Buffer. The pH in the cytosol of plant cells is around 7.5, and the homogenate should be buffered at pH 7–8. A buffer should therefore be included and a plethora of different buffers are available. Choose an inexpensive one with a pKa of 7–8 (e.g., 4-morpholinopropane sulfonic acid = MOPS) and be sure to use enough to counterbalance the acidic vacuolar content. The tissue type, the tissue/medium ratio, and the buffer concentration determine whether the acidic vacuolar content is properly neutralized. Obviously, special tissues like lemons or CAM plants harvested late at night or in the early morning, require special treatment!

Metal ion chelators. EDTA is usually included to bind mainly Ca^{2+} and Mg^{2+}, which may otherwise activate proteolytic enzymes.

Antioxidants. Antioxidants (reductants) such as ascorbic acid, dithiothreitol, or cysteine are used mainly to protect sulfhydryl groups on proteins against oxidation.

Bovine serum albumin (BSA). This protein from the blood of cows is usually included to protect the mitochondria: BSA binds fatty acids (its function in the blood) released by the action of phospholipases on phospholipids. BSA may also bind phenolics that are released from the vacuole. Both of these groups of compounds can otherwise be troublesome later. In addition, BSA may prevent proteases from degrading mitochondrial proteins by serving as an alternative substrate present at high concentrations.

Polyvinylpyrrolidone (PVP). A polymer that is often included, especially when using leaf material, to adsorb phenolic compounds.

Filtration

To remove larger tissue fragments, cell walls, and often starch grains, the homogenate is always filtered, usually through a nylon net with a mesh size of 30–100 μm.

Purification of Mitochondria

The first step of the purification is usually differential centrifugation. This means that the filtered homogenate is centrifuged first at low speed for a short time (e.g., 3000 g for 5–10 min.) and the pellet (containing cell wall fragments, starch grains, nuclei, intact plastids) is discarded. The supernatant is then centrifuged faster and for a longer time (e.g., 10,000 g for 10–20 min.) and the pellet (termed *crude mitochondria*) is collected. It contains intact and damaged mitochondria as well as peroxisomes and plastid membranes. The supernatant is discarded.

As soon as the mitochondria have been pelleted and thus removed from potentially damaging soluble substances in the homogenate, the medium can be greatly simplified. Usually the so-called wash medium contains only osmoticum and weak buffer (no need for a strong buffer since there is no vacuole content to neutralize). However, it may be wise to include EDTA to bind divalent cations as well as BSA to adsorb any fatty acids released by the action of phospholipases. The resuspended crude mitochondria are often diluted to 50–100 ml with wash medium and subjected to another round of slow and fast centrifugation to remove more contaminants.

The aim of the next step is to separate the intact mitochondria from the contaminants. This usually involves a density gradient centrifugation with Percoll either alone (e.g., Struglics et al. 1993) or together with other dense materials (e.g., Day et al. 1985). Percoll (colloidal silica coated with PVP) is inert and osmotically inactive. It has the further advantage that a Percoll gradient is self-generating because the Percoll microspheres pellet very slowly during the centrifugation. The crude mitochondria are layered on top of the wash medium containing the appropriate concentration of Percoll (usually 25–30%). If the Percoll concentration and the speed and duration of the centrifugation have been chosen correctly, the intact mitochondria will form a distinct band, well away from the contaminants. This band can be collected, washed free of Percoll by repeated dilution and centrifugation, and either used immediately, or frozen with 5% (v/v) dimethylsulfoxide (to preserve integrity) in liquid nitrogen and stored at -80°C until use. This preparation can be stable for months, retaining the ability to form a proton electrochemical gradient over the inner membrane.

The yield is 1–50 mg mitochondrial protein per 100 g fresh weight, depending on the tissue. This represents only a small fraction of the total amount of mitochondria in the tissue. The purification of mitochondria from green leaves presents special problems and a good protocol is described in Day et al. (1985). There is, at present, no good protocol for the isolation of both pure and functional mitochondria from Arabidopsis leaves (Hausmann et al. 2003; Keech et al. 2005), but a protocol for the purification of mitochondria from Arabidopsis cell cultures is given in Kruft et al. (2001). Finally, Boutry et al. (1984) demonstrate how purified mitochondria can be isolated from small amounts of plant tissue (down to 0.5 g fresh weight).

Mitochondria are rod-shaped organelles that can be considered the power generators of the cell, converting oxygen and nutrients into adenosine triphosphate (ATP). ATP is the chemical energy "currency" of the cell that powers the cell's metabolic activities.

This process is called aerobic respiration and is the reason animals breathe oxygen. Without mitochondria (singular, mitochondrion), higher animals would likely not exist because their cells would only be able to obtain energy from anaerobic respiration (in the absence of oxygen), a process much less efficient than aerobic respiration. In fact, mitochondria enable cells to produce 15 times more ATP than they could otherwise, and complex animals, like humans, need large amounts of energy in order to survive.

The number of mitochondria present in a cell depends upon the metabolic requirements of that cell, and may range from a single large mitochondrion to thousands of the organelles. Mitochondria, which are found in nearly all eukaryotes, including plants, animals, fungi, and protists, are large enough to be observed with a light microscope and were first discovered in the 1800s. The name of the organelles was coined to reflect the way they looked to the first scientists to observe them, stemming from the Greek words for "thread" and "granule." For many years after their discovery, mitochondria were commonly believed to transmit hereditary information. It was not until the mid-1950s when a method for isolating the organelles intact was developed that the modern understanding of mitochondrial function was worked out.

The elaborate structure of a mitochondrion is very important to the functioning of the. Two specialized membranes encircle each mitochondrion present in a cell, dividing the organelle into a narrow intermembrane space and a much larger internal matrix, each of which contains highly specialized proteins. The outer membrane of a mitochondrion contains many channels formed by the protein porin and acts like a sieve, filtering out molecules that are too big. Similarly, the inner membrane, which is highly convoluted so that a large number of infoldings called cristae are formed, also allows only certain molecules to pass through it and is much more selective than the outer membrane. To make certain that only those materials essential to the matrix are allowed into it, the inner membrane utilizes a group of transport proteins that will only transport the correct molecules. Together, the various compartments of a mitochondrion are able to work in harmony to generate ATP in a complex multi-step process.

Mitochondria are generally oblong organelles, which range in size between 1 and 10 micrometers in length, and occur in numbers that directly correlate with the cell's level of metabolic activity. The organelles are quite flexible, however, and time-lapse studies of living cells have demonstrated that mitochondria change shape rapidly and move about in the cell almost constantly. Movements of the organelles appear to be linked in some way to the microtubules present in the cell, and are probably transported along the network with motor proteins. Consequently, mitochondria may be organized into lengthy travelling chains, packed tightly into relatively stable groups, or appear in many other formations based upon the particular needs of the cell and the characteristics of its microtubular network.

The mitochondrion is different from most other organelles because it has its own circular DNA (similar to the DNA of prokaryotes) and reproduces independently of the cell in which it is found; an apparent case of endosymbiosis. Scientists hypothesize that millions of years ago small, free-living prokaryotes were engulfed, but not consumed, by larger prokaryotes, perhaps because they were able to resist the digestive enzymes of the host organism. The two organisms developed a symbiotic relationship over time, the larger organism providing the smaller with ample nutrients and the smaller organism providing ATP molecules to the larger one. Eventually, according to this view, the larger organism developed into the eukaryotic cell and the smaller organism into the mitochondrion.

Mitochondrial DNA is localized to the matrix, which also contains a host of enzymes, as well as ribosomes for protein synthesis. Many of the critical metabolic steps of cellular respiration are catalysed by enzymes that are able to diffuse through the mitochondrial matrix. The other proteins involved in respiration, including the enzyme that generates ATP, are embedded within the mitochondrial inner membrane. Infolding of the cristae dramatically increases the surface area available for hosting the enzymes responsible for cellular respiration.

Methods for Establishing Purity, Intactness, and Functionality of the Fractions

It is often important to determine the degree of purity of the mitochondria and their subfractions.

The intactness of the mitochondria is usually established with enzyme assays that take advantage of the fact that an intact membrane is not permeable to certain molecules.

For instance, the outer mitochondrial membrane is impermeable to cytochrome *c*, a 12.5 kDa protein that is too large to pass through the outer membrane pores, and NADH cannot pass through an intact inner membrane.

Functionality is usually tested using an oxygen electrode. If the mitochondria show high rates of oxygen consumption, high respiratory control ratios (>3), and ADP:O ratios close to the theoretical values they are in very good shape. Note that, for mitochondria with an appreciable alternative oxidase activity (measured in the presence of the cytochrome oxidase inhibitor, KCN), the respiratory control ratio and ADP:O ratio should be measured in the presence of an alternative oxidase inhibitor such as salicylhydroxamic acid.

Mitochondrial Activities can be Determined in Situ

It is relatively straightforward to measure the activity of enzymes in the soluble matrix fraction isolated from mitochondria.

However, this fraction is very diluted compared to the matrix inside intact mitochondria (300–500 mg protein/ml!) and protein–protein interactions may have been lost. When measuring the activity of enzymes in fractions containing membranes—intact mitochondria and submitochondrial particles—a detergent can often be used to remove the permeability barrier that the inner membrane constitutes.

However, it is also possible to use the ionophore alamethicin to permeabilize the inner membrane to molecules <1000 Da, which makes it possible to measure the activity of matrix and inner membrane enzymes in their native proteinaceous environment.

Using alamethicin, it is even possible to measure the activity of cytosolic and mitochondrial enzymes inside intact cells since the ionophore also permeabilizes the plasma membrane.

The Q-Cycle Explains How Protons Can Be Pumped across the Inner Mitochondrial Membrane

The electron transport chain (ETC) in the mitochondrial inner membrane of most eukaryotes consists of four complexes, complex I to complex IV. Complexes I, III and IV pump protons over the inner mitochondrial membrane when transporting electrons.

For each electron pair (corresponding to one NADH or one oxygen atom), four protons are pumped by complex I, four by complex III, and two by complex IV. The detailed three-dimensional structures have been determined for the mammalian complexes III and IV, but not for complex I. The structural information has improved the understanding of the electron paths of complexes III and IV, but the proton-pumping function of the latter is still not fully understood. For complex I, neither electron nor proton paths are known in any detail.

Ubiquinone (Q), also called coenzyme Q, is a mobile electron and proton carrier in the inner mitochondrial membrane, structurally and functionally similar to plastoquinone in the thylakoid membrane of plastids (Figure). The ability of Q to carry both electrons and protons—and that it can accept and donate them step-wise—is utilized in electron transport and proton pumping (Figure) (Nicholls and Ferguson 2002).

To be reduced, it can receive two electrons from complex I or succinate dehydrogenase and take up two protons from the matrix, forming QH_2. However, at the oxidizing side (complex III), there is actually both oxidation and reduction taking place. A QH_2 will be oxidized to Q at the outer Q-site, releasing two electrons. One of these electrons goes to cytochrome *c*, whereas the other is recycled to reduce an oxidized Q in the inner Q-site—one step to becoming a semiquinone.

The next time a QH_2 is oxidized at the outer Q-site, the semiquinone in the inner site will receive a second electron, becoming a QH_2. So, from two QH_2, we have made one Q and one QH_2, which again can be oxidized. In each cycle (i.e., for each electron passing to cytochrome *c*), two protons are taken up from the matrix and two protons released to the intermembrane space. So, the Q-cycle mediates the pumping of an extra proton per electron transported, increasing energy conservation.

If we look at the whole reaction from succinate to cytochrome *c*, we see that for each succinate, four protons are transported from the matrix to the intermembrane space.

However, at the same time, the two electrons are also transported out. So, this corresponds to the net pumping out of two positively charged species. Since, energetically, the mitochondrial membrane potential (which determines the energetic cost of transporting a positive charge out) is much larger than the pH gradient (which determines the cost for electroneutral export of a proton), the complex III proton pumping is less energy-demanding than, for example, complex I, which pumps four protons in their completeness.

Multiple Energy-Conservation Bypasses in Oxidative Phosphorylation of Plant Mitochondria

The electron transport chain (ETC) in the mitochondrial inner membrane of most eukaryotes consists of four large protein complexes, complex I to complex IV. The electron transport activity of complexes I, III and IV (the latter two, often jointly called the cytochrome pathway) is coupled to extrusion of protons across the inner mitochondria.

The electrochemical proton gradient that is formed and maintained in the process is used by the ATP synthase (also called complex V) to make ATP (Figure). The plant ETC, however, is highly branched. As a consequence, the plant oxidative phosphorylation system has several alternative pathways of electron transport via type II NAD(P)H dehydrogenases and the alternative oxidase (AOX), and additionally, an alternative proton transporter called the uncoupling protein (UCP).

These proteins mediate bypasses around the proton-translocating multi-protein complexes (I–V): type II NAD(P)H dehydrogenases bypass complex I, AOX bypasses complexes III and IV, and UCP bypasses the ATP synthase (Siedow and Umbach 1995; Vanlerberghe and McIntosh 1997; Møller 2001; Vercesi 2001; Millenaar and Lambers 2003; Fernie et al. 2004; Rasmusson et al. 2004).

The type II NAD(P)H dehydrogenases and the AOX transport electrons without pumping protons, whereas the UCP allows proton flow from the intermembrane space to the matrix without ATP synthesis. The consequence of this is that flux through the energy conservation bypasses will neither contribute to respiratory ATP production, nor will they be controlled by cellular adenylate status. The special features of these energy bypass proteins and discuss the consequences of their existence to the plant.

NAD(P)H Dehydrogenases in Plant Mitochondria

Except for yeast, the ETC of all eukaryotes—including plants—contains a proton-pumping NADH dehydrogenase, complex I, which is inhibited specifically by rotenone. Additionally, plant, fungal and protist mitochondria contain different setups of rotenone-insensitive enzymes capable of oxidizing NADH and/or NAD(P)H and passing the electrons on to ubiquinone.

These NAD(P)H dehydrogenases are classified as type II, whereas complex I belongs to type I. The inner mitochondrial membrane in plants contains such NAD(P)H dehydrogenases, facing either the matrix or the intermembrane space, thus oxidizing NAD(P)H either from the matrix or the cytoplasm.

The outer membrane of all mitochondria contains NADH-ferricyanide and NADH-cytochrome *b* reductase activities that are insensitive to rotenone and to the complex III inhibitor, antimycin A.

There are Three NAD(P)H Dehydrogenases on the Inner Matrix Surface of the Inner Membrane

Much of the information about the number of matrix-facing NAD(P)H dehydrogenases has come from experiments using inside-out inner membrane vesicles. Such vesicles have the matrix surface facing the medium and the external NAD(P)H dehydrogenases hidden inside the vesicles and unable to interfere with the assays.

Recently, this method has been complemented with a technique where one uses a peptide called alamethicin to make a channel for the substrate through the inner membrane. Experiments with these methods show the presence of three NAD(P)H dehydrogenase activities on the matrix side of the inner membrane:

- Complex I pumps protons, is Ca^{2+}-independent and is inhibited by rotenone and diphenylene iodonium (DPI). Complex I has a high affinity for NADH.
- ND_{in}(NADH) does not pump protons, is Ca^{2+}-independent and DPI insensitive. The enzyme has a tenfold lower affinity for NADH than complex I.
- ND_{in}(NADPH) does not pump protons, is Ca^{2+}-dependent and very sensitive to DPI.

Because of the contrasting affinities of complex I and ND_{in}(NADH) for NADH, only complex I is engaged at low matrix concentrations of NADH. However, when the NADH concentration in the matrix increases (e.g., if the ETC is restricted by lack of ADP in the cell), ND_{in}(NADH) can become engaged. When electrons are transported through the type II NAD(P)H dehydrogenases, only the two proton pumping sites of the cytochrome pathway are involved in the flow of electrons from matrix NADH to oxygen. As a consequence, the theoretical ADP/O yield is 1.5 unless the AOX or the UCP are active, in which case it will be even lower.

The External NAD(P)H Dehydrogenases are Ca^{2+}-Dependent

Plant mitochondria can oxidize cytosolic NADH and NAD(P)H directly via external dehydrogenases. The NADH and NAD(P)H dehydrogenase activities are catalyzed by separate enzymes. None of the external dehydrogenases pumps protons, which means that, at most, two sites of proton-pumping are involved when the electrons move from the dehydrogenases to oxygen. So, as for ND_{in}(NADH), the theoretical maximum ADP/O yield is 1.5.

The oxidation of exogenous NADH or NAD(P)H by most plant mitochondria requires Ca^{2+}, and as a result, is inhibited by EDTA and EGTA, chelators of di-and trivalent cations. *In vitro*, the Ca^{2+}concentration needed for half-maximal activity of NADH oxidation is 0.1–0.5 μM, depending on ionic environment and respiratory state This is in the same range as the 0.1–0.2 μM free Ca^{2+} thought to be present in the cytosol of a resting plant cell. Therefore, the external NAD(P)H oxidation may become activated by the increased Ca^{2+}-concentration associated with cellular excitation due to external conditions.

Mammalian mitochondria lack external NAD(P)H dehydrogenases. Instead, they have a Ca^{2+}-dependent glycerol-3-phosphate dehydrogenase on the outer surface of the inner membrane, which donates electrons directly to the ubiquinone pool.

The dihydroxyacetone phosphate formed is converted back into glycerol-3-phosphate by a cytosolic NAD-dependent glycerol-3-phosphate dehydrogenase. In this way, the two low-molecular-weight compounds, glycerol-3-phosphate and dihydroxyacetone phosphate, shuttle redox equivalents from the cytosol, across the outer membrane and to the inner membrane to yield the same amount of ATP as is produced by the plant mechanism. Yeast have both external NADH dehydrogenases and a glycerol-3-phosphate shuttle, and since a mitochondrial glycerol-3-phosphate dehydrogenase has been found in plant mitochondria, plants may also have both systems.

Another way to oxidize cytosolic NAD(P)H that is found in mammalian mitochondria is to use the malate/aspartate shuttle to tranfer reductant (a collective term for high-energy electrons carried by, for example, NAD(P)H, ferredoxin or an organic acid redox couple like malate/oxaloacetate) into the matrix as NADH that can be oxidized by the internal NAD(P)H dehydrogenases. However, it is not clear whether this shuttle can work for import of cytosolic reductant into plant mitochondria.

In the opposite orientation, NADH is shuttled from the matrix to the cytosol via the malate/oxaloacetate shuttle. In this way, the external NADH dehydrogenase can also work as a bypass around complex I. Support for this bypass has been seen in the form of an upregulation of external NADH dehydrogenase activity in mutants where complex I or an internal type II NAD(P)H dehydrogenase has been inactivated.

The Molecular Identities of the Rotenone-Insensitive NAD(P)H Dehydrogenases

The type II NADH dehydrogenases from *E. coli* and yeast are small, single polypeptide proteins of around 50 kDa. They oxidize only NADH and contain FAD as an electron-accepting prosthetic group.

Attempts to purify the type II NAD(P)H dehydrogenases have been unsuccessful in identifying the plant proteins. However, the characterization of potato homologues to the type II NADH dehydrogenases in *E. coli* and yeasts have provided candidate plant proteins to try and assign to the different reported activities.

One protein, NDA1, is internal and the other, NDB1, external of the inner mitochondrial membrane. The latter was shown to be an external calcium-dependent NAD(P)H dehydrogenase. Like a related fungal NAD(P)H dehydrogenase, NDB1 contains a motif for Ca-binding, indicating that in the activation of plant and fungal NAD(P)H dehydrogenase by Ca^{2+}, the ion binds directly to the enzyme. In Arabidopsis, a weed for which the nuclear genome sequence is known, small families of both NDA-type and NDB-type proteins are encoded by nuclear genes. One of the NDA-type proteins has been shown to be an internal NAD(P)H dehydrogenase. Additionally, a third type of type II NAD(P)H dehydrogenase was found in Arabidopsis, the NDC1. This protein is most closely related to cyanobacterial enzymes, suggesting that the gene entered the eukaryotic domain with the plastid, later to be transferred to the nucleus.

The Alternative Oxidase

The presence of the AOX is one of the features that sets plant (and fungal) mitochondria apart from mammalian mitochondria. This enzyme has therefore received a lot of attention from plant scientists and we now have a fairly good understanding of the regulation of its structure and activity.

The AOX is a quinol-oxygen oxidoreductase and it does not pump protons. The oxygenase transfers electrons from ubiquinol (the reduced form of ubiquinone) to oxygen and generates water as the end-product of the reaction, so four electrons must be transferred to oxygen.

The precise mechanism is not known, but the protein has been shown to contain a catalytical di-iron centre. The functional form of the enzyme is a dimer with the two polypeptides either covalently or non-covalently bound to each other (Figure). The predicted structure places the protein embedded into the inner leaflet of the inner mitochondrial membrane, with the active site and the regulatory cysteines exposed to the matrix.

When electrons from NADH oxidation flow through the AOX, at least two of the three sites of proton translocation are bypassed, so energy conservation in the form of ATP is much smaller when the AOX is active. The activity of the AOX must therefore be carefully regulated. There are often several AOX isoforms in each plant species—for example, five genes encoding AOX proteins are present in Arabidopsis. The relative amounts of three isoforms of AOX vary with development, and cellular redox status, as well as external factors.

The AOX is present in the inner membrane as a dimer in which the two subunits can be covalently linked through a disulfide bridge between cysteine residues. The reduced non-covalently linked dimer is the active form, and this form appears to be the predominant one *in vivo*. It was recently shown that a mitochondrial thioredoxin system regulates AOX. Thioredoxins regulate enzymes by reducing disulfide bridges, and it is likely that the regulation involves the redox-active cysteines.

The presence of thioredoxin regulation connects the AOX activation level to the mitochondrial NAD(P)H pool and the oxidation of citric acid cycle metabolites by NADP-utilizing enzymes.

The activity of AOX is also stimulated by 2-oxo organic acids—in particular, pyruvate, which activates by interacting with the same cysteine that is involved in dimerization. This activation may be a feed-forward mechanism that upregulates the capacity of the ETC when the glycolytic end-product is available.

Both the cytochrome pathway and the AOX pathway use ubiquinol as the substrate. In the absence of pyruvate, the cytochrome pathway is saturated at 40% Q_{red}/Q_{tot} (40% of the UQ is reduced), at which point the AOX pathway is not engaged at all. The AOX pathway starts accepting electrons at 50% reduction and is not fully saturated at 80%.

In the presence of pyruvate, the AOX becomes engaged at a much lower Q_{red}/Q_{tot} (20%), whereas the cytochrome pathway is unaffected. It is puzzling, however, that the normal *in vivo* concentration of pyruvate appears to be higher than the concentration of pyruvate required to give maximal stimulation of AOX activity *in vitro*. It thus appears that there is little regulation of AOX activity via pyruvate under physiological conditions.

AOX Activity Can Be Measured In Vivo

For many years, benzhydroxamic acid derivatives,, were used to quantify the capacity and activity of the AOX pathway in both isolated mitochondria and intact tissues. AOX capacity was determined from the rate of respiration in the presence of KCN (which blocks the cytochrome pathway), and corrected for residual respiration in the presence of both KCN and SHAM.

AOX *in vivo* activity was determined from the decrease in the rate of respiration when SHAM was added in the absence of KCN. The assumption was that the cytochrome pathway was always used first and that the AOX would be engaged only when the cytochrome pathway was saturated. However, as we have seen above, this assumption is not correct. The consequence of this is that *in vivo* activity of AOX cannot be measured using inhibitors alone.

A better method for measuring AOX *in vivo* activity is now available. It makes use of the fact that cytochrome *c* oxidase and AOX discriminates differently between ^{18}O relative to ^{16}O.

The so-called oxygen discrimination of the sample (tissue or mitochondria) can be measured by mass spectrometry: (1) with KCN to resolve the oxygen discrimination by AOX; (2) with SHAM to resolve the oxygen discrimination by cytochrome *c* oxidase; and (3) without inhibitors to resolve the discrimination value resulting from the activity of both enzymes.

From these values, the percentage contribution by AOX and cytochrome *c* oxidase to the total uninhibited rate of respiration can be calculated. An example of the use of the oxygen discrimination technique with an intact tissue is given below. A remaining problem in the estimation of AOX activity *in vivo* is that measurements can only be made in darkness (to avoid interference from photosynthetic oxygen exchange).

Plant Mitochondria also Contain an Uncoupling Protein

Mammalian mitochondria contain neither type II NAD(P)H dehydrogenases nor AOX. However, they have a UCP that increases the proton permeability of the inner mitochondrial membrane and, in that way, dissipates the proton gradient. This is another mechanism for reducing the ATP production and increasing heat production, for example, for thermoregulation.

Surprisingly, plant mitochondria also contain a UCP. We do not know for certain why plant mitochondria should require two different mechanisms for achieving the same end result. The UCP is related to anion carriers. Its *in vitro* transport activity is dependent on free fatty acids and is inhibited by nucleotides like ATP and GTP but stimulated by superoxide. The latter result indicates that the UCP may protect against oxidative stress.

The Physiological Impact of Energy Bypass Proteins

The respiratory rates of some thermogenic flowers are the highest among plants and, in fact, exceed even those of warm-blooded animals. The external NADH dehydrogenase, AOX and UCP appear particularly active in thermogenic flowers where they are primarily responsible for heat production. At the same time it is clear that the respiratory rate of most plant tissues is insufficient to generate an appreciable increase in tissue temperature.

AOX is up-regulated by cold stress, although this is clearly not to heat the tissue. However, during cold stress, reactive oxygen species (ROS) are formed, and also other stress treatments that lead to elevated ROS will induce AOX. ROS (e.g., superoxide and hydrogen peroxide) are formed as by-products of electron transport under aerobic conditions.

ROS can cause damage to proteins, lipids, and DNA and the cell must therefore limit their formation. AOX helps prevent overreduction of the ETC and thus lowers ROS production. Supporting this, Maxwell et al. (1999) demonstrated the importance of the AOX for limiting ROS formation in living plant cells.

Overexpression of AOX lowered the steady state level of ROS, whereas suppression increased it. This result also points to the mitochondrion as an important site of ROS generation in plant cells. With the recent discovery of ROS stimulation of plant UCP and the increased ROS resistance in plants overexpressing UCP, the picture has become more complex. By decreasing the resistance for electron transport through the proton-pumping complexes, UCP activity may, like AOX, lead to decreased overreduction and ROS formation. Another factor speaking in favour of UCP having a role in ROS protection is that the AOX is inhibited by an ROS product.

Both type II NAD(P)H dehydrogenases and AOX may act as reductant overflow enzymes by reoxidizing surplus NADH to oxygen. This can prevent inhibition of the citric acid cycle under conditions in which the ADP concentration is low, and thus the proton-pumping complexes are restricted by a high membrane potential. In this way, the citric acid cycle can supply carbon skeletons for biosynthesis, even if the ATP level in the cell is high.

The concept of overflow is also relevant for photorespiratory conditions, under which there is a massive flow of glycine that has to be oxidized by the mitochondria. As much as half of the NADH produced in this process is probably exported, but the other half has to be oxidized promptly. There are several reasons to believe that a reductant overflow via energy bypass proteins is involved.

The ND_{in}(NADH) and AOX capacity in leaf mitochondria is higher than in mitochondria from heterotrophic tissue, and both proteins are down-regulated by dark treatment. Consistent with these observations, AOX activity *in vivo* is up-regulated by light (Ribas-Carbo et al. 2000), AOX capacity is induced by greening of etiolated tissue, and expression of the *nda*1 gene [encoding an internal NAD(P)H dehydrogenase] in potato and Arabidopsis is light-dependent.

Mutant tobacco plants, lacking the proton-pumping complex I can still survive, apparently aided by the action of ND_{in}(NADH), which oxidizes matrix NADH with a lower ATP yield.

These plants, however, show decreased photosynthetic rates, indicating that the ND_{in}(NADH) cannot fully compensate for the missing complex I. Also, external NAD(P)H dehydrogenases have been suggested to support photosynthesis. These enzymes may reoxidise NAD(P)H that is shuttled out of the chloroplast in order to balance the NAD(P)H to ATP ratio in the stroma. In summary, there are many indications that mitochondria—especially the energy bypass proteins—have roles in supporting photosynthesis, for example by oxidizing excess NAD(P)H.

The AOX pathway is involved in the acidification/deacidification cycle of CAM plants. CAM plants fix CO_2 in the form of malate during the night. This is done by cytosolic PEP carboxylase, and the malate produced is stored in the vacuole.

The CO_2 fixed is released during the daytime by malic enzymes, located in cytosol and mitochondria. Robinson et al. (1992) used the oxygen discrimination technique to measure the contribution of the AOX pathway to whole-leaf respiration in the CAM-plant, *Kalanchoë daigremontiana.*

They found that the respiratory rate of the leaves increased by 35% in the early hours of the day. This increase was due to increased AOX activity. A more detailed time course showed that the AOX activity increased two-to threefold during the early day at the same time as the malate was metabolized. As soon as the malate had disappeared by mid-day, the AOX activity returned to the low level observed during the night. A reasonable conclusion is that a substantial amount of the malate from the vacuole is oxidized via malic enzyme in the mitochondria. This generates NADH in the matrix, which is oxidized to a large extent via the AOX. The released CO_2 can then be refixed by the Calvin cycle in the chloroplast.

F_oF_1-ATP Synthases: The Worlds Smallest Rotary Motors

F_oF_1-ATP synthases (also called F-ATPases) are present in the inner membranes of mitochondria, chloroplasts, and bacteria. These large, multisubunit enzymes consist of a water-soluble catalytic complex (F_1)

attached to an integral membrane protein complex (F_o) that transports protons across the membrane. The F_1 complex is composed of at least five different types of subunits (three α, three β, one γ, one δ, and one å).

When the F_1 complex is dissociated from the membrane, it is active as an ATPase. In fact, under the appropriate conditions the intact F_oF_1-ATP synthase can run in reverse and act as a proton pump, using the energy of ATP hydrolysis to move H^+ across the membrane. Our understanding of how ATP is synthesized was advanced by Paul Boyer at the University of California, Los Angeles, who proposed the binding-change mechanism for catalysis by F-ATPases. From the detailed understanding of how the ATPases function, he proposed that the binding-change mechanism for ATP synthesis contains three important components:

1. The major energy-requiring step is not the synthesis of ATP from ADP and P_i, but the release of ATP from the enzyme.
2. Substrate is bound and products are released at three separate but interacting catalytic sites, corresponding to the three catalytic subunits (b subunits). Each catalytic site can exist in one of three conformations: tight, loose, or open.
3. The binding changes are coupled to proton transport by rotation of the g subunit. That is, the flow of protons down their electrochemical gradient through the F_o complex causes the g subunit to rotate.

Rotation of the γ subunit then brings about the conformational changes in the catalytic complex that allow the release of ATP from the enzyme, and the reaction is driven forward. The reverse occurs when the enzyme functions as an ATP-driven proton pump.

The first two predictions of the binding-change model are supported by many lines of evidence, mainly kinetic studies, and are now generally accepted. However, the prediction of a rotary mechanism for coupling proton flow to ATP synthesis has been more difficult to demonstrate. Two major breakthroughs have led to confirmation of the third prediction, as well.

The first breakthrough was the determination of the crystal structure of the F_1-part of bovine mitochondrial ATPase by the laboratory of John Walker in Cambridge, England. The crystal structure showed that the three catalytic β subunits differ in their conformations and in the nucleotide bound to them, consistent with the binding-change mechanism.

Even more exciting was the discovery that the γ subunit is inserted like a shaft through the centre of the catalytic complex, which consists of three α subunits and three β subunits arranged alternately in a doughnutlike structure. Moreover, the interface between the γ subunit and the α and β subunits is highly hydrophobic.

The hydrophobicity of the interface minimizes the interactions between the subunits, consistent with the rotation of the γ subunit within the hole formed by the catalytic complex. In other words, the γ subunit looks like a molecular bearing lubricated by a hydrophobic interface. Although many questions were answered by elucidation of the crystal structure of the F_1-ATP synthase, the rotational model cannot be tested by a static "snapshot" of the enzyme. Definitive demonstration of rotation requires a video recording of the spinning of the enzyme in real time. But although they are large for proteins, F_1-ATP synthases are still far too small to be visualized in a light microscope.

To visualize the rotation of the enzyme, Masasuke Yoshida and his colleagues at the Tokyo Institute of Technology came up with an ingenious method to make the enzyme much larger than it is. As Figure below shows, they attached an actin filament labelled with a fluorescent dye to the base of the γ subunit using another protein as a "glue." They then attached the F_1 complex upside down to a glass surface.

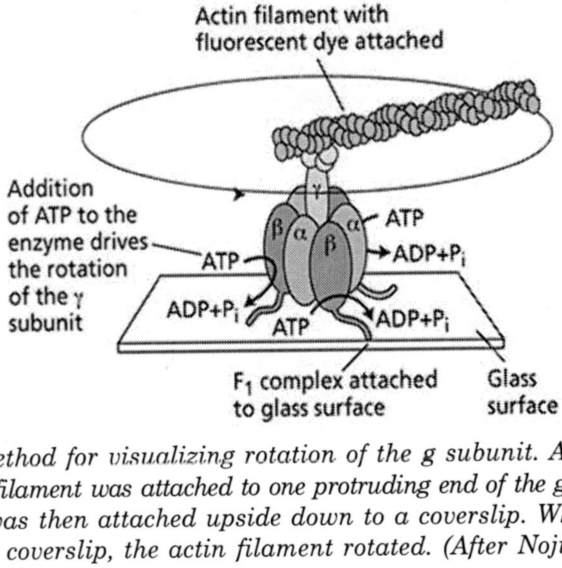

Figure: *A method for visualizing rotation of the g subunit. A fluorescently labelled actin filament was attached to one protruding end of the g subunit. The F_1 complex was then attached upside down to a coverslip. When ATP was added to the coverslip, the actin filament rotated. (After Noji et al. 1997.)*

If the γ subunit rotates with respect to the catalytic complex, the actin filament should swing around with it. Since the filament is very long compared to the ATP synthase (about 1 μm), its rotation should be visible in a fluorescence microscope. In other words, the fluorescently tagged actin filament, which is large enough to visualize in a light microscope, reports the rotation of the γ subunit.

The results were spectacular! When ATP was added to the modified enzyme, the actin filaments were seen to swing around in a circle at as much as 4 revolutions per second in a fluorescence microscope. To give some idea of scale, if you were a γ subunit, this rotation rate would be equivalent to swinging a several hundred meter-long rod around your head at 4 revolutions per second in water! However, the measured velocity is undoubtedly an underestimation of the actual velocity in vivo, because of the enormous torque required to swing such a large mass. Demonstration of the rotary motion of the γ subunit made it possible to put together a model of how the ATP synthase works.

For their contributions to elucidation of the mechanism of ATP synthesis, Paul Boyer and John Walker shared half the Nobel prize in physiology or medicine in 1997. The other half went to Jens Skou for his pioneering work on the K^+,Na^+-ATP synthase, the mammalian counterpart of the plant plasma membrane H^+-ATP synthase.

Transport Into and Out of Plant Mitochondria

Plant mitochondria must exchange metabolites and information (signal transduction) with the cytosol. To give some central examples, the end products of glycolysis, pyruvate and malate, need to be taken up, and citric acid cycle intermediates exported to be used as building blocks in biosynthetic processes.

Likewise, ADP and P_i must be taken up and ATP exported for the mitochondrion to fulfil its role as the powerhouse of the cell. To carry out the transport of this large number of molecules, plant mitochondria possess a family of mitochondrial carrier proteins.

These are transmembrane proteins with a common general structure and they typically transport relatively small charged molecules, like metabolites. In Arabidopsis there are almost 60 members, most of which are not functionally characterized, giving a hint to the complexity of the mitochondrial transport processes. However, mitochondria exchange more than metabolites, and therefore also other types of transporters are involved.

The outer membrane is permeable to molecules of less than 10 kDa, (i.e., all non-polymeric compounds). The permeability results from the presence of pores, which appear to be open in most *in vitro* experiments. Recent studies in animals have characterized pore-forming proteins in the mitochondrial outer membrane, called porins, which appear to regulate transport across the membrane.

Plant mitochondria also have porins but their characterization awaits further research. In our discussion here we will focus on the regulation of mitochondrial transport across the inner mitochondrial membrane.

We will first consider the relatively few molecules that can move across the inner membrane without the help of a specific carrier protein. We will then discuss the various metabolite transporters and, finally, briefly look at the transport of proteins and RNA.

Small, Neutral, and Hydrophobic Molecules Diffuse Across the Inner Membrane

Irrespective of size, molecules that either are charged or are very hydrophilic can only cross a lipid bilayer with the help of a specific protein carrier.

Small, *neutral* molecules can diffuse across the lipid bilayer and do not require protein transporters. Among those are O_2 and CO_2, two important molecules for respiration. O_2 is needed on the inner surface of the inner membrane where the active sites of both cytochrome *c* oxidase and the alternative oxidase are located.

The CO_2, produced as an end product in the citric acid cycle, needs to diffuse out of the mitochondrion, ultimately to be lost to the surroundings or re-fixed in the chloroplasts. In addition, however, it is possible that CO_2 can also be transported out of the mitochondrion as bicarbonate (HCO_3^-). Mitochondria are osmotically active and swell in a hypotonic medium. This means that the inner membrane is also permeable to H_2O, but it is not clear whether this occurs by diffusion or if aquaporins may be involved.

For technical reasons, it is of importance that several well-known inhibitors of mitochondrial function can diffuse into mitochondria because they are uncharged and hydrophobic. These are rotenone, the specific inhibitor of complex I, antimycin, the specific inhibitor of complex III, HCN, the protonated form of the cyanide (CN^-) inhibitor of complex IV (and other heme-containing enzymes), and oligomycin, the specific inhibitor of the F_oF_1-ATP synthase.

Transport of Inorganic Ions

Inorganic phosphate ($H_2PO_4^-$) is taken up into the matrix in exchange for a hydroxyl ion (HO^-). The energy for the uptake is supplied by the proton gradient (–pH). We will see below that the phosphate gradient, in turn, drives transport of other metabolites.

Several other inorganic ions—Mg^{2+}, Ca^{2+}, K^+, and NH_4^+—are involved in mitochondrial metabolism, but the mechanism by which they pass the inner membrane of plant mitochondria is not well understood. In mammalian mitochondria, uptake of K^+ through a K^+-specific channel, and extrusion via a K^+/H^+ antiporter, is involved in matrix volume regulation and cellular signalling (Garlid and Paucek 2003). A similar system of K^+ transport is present in plant mitochondria.

Both mammalian and plant mitochondria contain an uncoupling protein (UCP). This protein facilitates the movement of protons across the inner membrane and therefore uncouples electron transport and decreases ATP yield.

Transport of Adenylates

One of the important functions of mitochondria is to supply the cell with ATP. To do this, mitochondria have an adenine nucleotide transporter that facilitates the exchange of ADP^{3-} (inward) for ATP^{4-} (outward). This exchange is driven by the membrane potential (inside negative), since there is a net outward movement of one negative charge.

In combination with the phosphate carrier reaction, this means that the uptake of one ADP and one phosphate and the export of one ATP (i.e., the transport event connected to mitochondrial ATP synthesis) costs one vectorial proton—the electrical component for adenylate exchange and the "pH component for phosphate uptake.

Transport of Citric Acid Cycle Intermediates and Related Compounds

Pyruvate and malate, the end products of glycolysis, both have transporters in the inner membrane. Pyruvate is taken up in exchange for a hydroxyl ion, and malate^{2-} in exchange for $H_2PO_4^{2-}$.

The transporter for malate also takes succinate^{2-} and is therefore called the dicarboxylate transporter. Malate^{2-} or succinate^{2-} can, in turn, exchange for citrate^{2-} on the tricarboxylate transporter. Malate^{2-} can also exchange for 2-oxoglutarate^{2-} via a special transporter. An interesting transporter, the oxaloacetate transporter, exchanges oxaloacetate^{2-} for malate^{2-} (or citrate^{2-}), which permits the shuttling of reducing equivalents.

The joint activity of these transporters allows plant mitochondria to import or export most citric acid intermediates. Finally, the glutamate/aspartate transporter facilitates the uptake of glutamate. In mammalian mitochondria, glutamate is taken up together with a proton, and it is thus driven by both the electrical and the chemical proton gradient. A similar transporter is present in plant mitochondria, but a proton co-transport has not been shown, so the transport may be passive in plants. Recently, several additional transporters for metabolites have been identified, and it is likely that certain setups of transporters are only expressed under special developmental conditions, like the mobilization of storage nutrients in developing seedlings.

In several cases the involvement of the mitochondrion in cellular metabolism implies that certain metabolites are taken up across the inner membrane, but the transporter has not yet been described. This is the case for the massive flux of glycine into—and CO_2, serine and NH_4^+ out of—mitochondria in C_3 leaves during photorespiration This is also the case for proline, which accumulates to high concentrations in the cytosol during episodes of plant water stress, and it is then broken down during the rehydration phase by a pathway involving two mitochondrial enzymes.

Transport of Coenzymes

Coenzymes are either synthesized inside the mitochondrion or imported. There is some evidence that, in addition to ATP, at least NAD^+, thiamine pyrophosphate, and coenzyme A are imported across the inner membrane.

The final step in ascorbic acid biosynthesis has recently been shown to be on the outer surface of the inner mitochondrial membrane. In view of the accumulating evidence that ascorbate has an important function in removing ROS inside the mitochondrion, we should expect to find an ascorbate transporter in the inner mitochondrial membrane.

The situation with folate (involved in 1C transfers) is the opposite to that with ascorbate: The only site of synthesis of many of the folate metabolites in the plant cell is the mitochondrial matrix. We can therefore expect that these molecules are exported via specific transporters.

Transport of Reducing Equivalents—Metabolic Shuttles

Mammalian mitochondria cannot oxidize cytosolic NADH or NAD(P)H directly, so they make use of "shuttle mechanisms." An exchange of metabolites by inner membrane transporters takes place through these shuttles so that a reduced compound moves into the matrix in exchange for a more oxidized compound moving out into the cytosol.

In this way, reducing equivalents are transported into the matrix ultimately to be oxidized by the respiratory chain. The shuttles can also be used to transport reducing equivalents in the opposite direction. Although plant mitochondria can oxidize cytosolic NADH and NAD(P)H directly by the two external NAD(P)H dehydrogenases, they can also use these two metabolic shuttles. The malate/oxaloacetate shuttle uses malate dehydrogenase in the cytosol and in the matrix to catalyze the interconversion of malate (reduced) and oxaloacetate (oxidized).

These two compounds are exchanged on the oxaloacetate transporter. This exchange, however, is not driven by the electrochemical proton gradient across the inner membrane, so it can only move reducing equivalents from a relatively reduced compartment to a relatively oxidized compartment. Since the mitochondrial matrix is more reduced than the cytosol, the oxaloacetate transporter is likely to work primarily in the export of reducing equivalents.

Under photorespiratory conditions, large amounts of glycine are oxidized in the mitochondria to give NADH while, at the same time, NADH is required in the peroxisomes to reduce hydroxypyruvate to glycerate. It is thought that as much as 50% of the matrix NADH is exported, probably via the malate/oxaloacetate shuttle.

The more complex malate/aspartate shuttle involves two matrix enzymes and two cytosolic enzymes as well as two inner membrane transporters. In mammals, it is powered by the electrochemical proton gradient in the direction of the import of reducing equivalents. However, since the glutamate/aspartate exchanger may be passive in plants, this shuttle in plants may similarly be independent of the electrochemical proton gradient. If so, it would (like the malate/oxaloacetate shuttle described above) only be able to function as an exporter of mitochondrial reductant.

Transport of Proteins—Protein Import

Plant mitochondria contain their own DNA (mtDNA) and ribosomes and are capable of synthesizing the proteins encoded in the mtDNA. However, plant mtDNA encodes only a few of the proteins found in the mitochondrion. The great majority of the mitochondrial proteins have to be imported from the cytosol where they are synthesized on free ribosomes. These are synthesized with a presequence or targeting peptide in the N-terminal end, which is recognized by receptors on the outer mitochondrial membrane.

The import of proteins requires a complex machinery with two multisubunit protein complexes—one in the outer membrane (TOM, transporter of the outer membrane) and one in the inner membrane (TIM, transporter of the inner membrane). Concomitant with, or immediately afterwards the import of the protein into the correct mitochondrial compartment, the targeting peptide is often cleaved off to yield the mature and active protein.

Transport of RNA

Mammalian mtDNA encodes the full complement of tRNA required for mitochondrial protein synthesis. In contrast, plant mtDNA only encodes some of the tRNAs required for protein synthesis within the mitochondrion. The missing tRNA must be imported by a yet-to-be-elucidated import mechanism.

The Genetic System in Plant Mitochondria has Some Special Features

As in plastids, the genetic system in mitochondria differs both structurally and functionally from the nuclear genome and its gene expression. On the other hand, mitochondria share several genetic features with bacteria.

For example, several genes are transcribed together and the ribosomes are of a prokaryotic type. This is in line with the evolutionary origin of the mitochondrion as an endosymbiont of the a-subclass of Proteobacteria, a group including the intracellular pathogen *Ricketsia prowazekii*.

Plants are also distinctly different from other eukaryotes in their mitochondrial DNA (mtDNA) and its expression. For example, the plant mtDNA is very large and highly variable in size between species. The presence of several repeats enables the genome to recombine, changing the gene order or even causing a split into smaller subgenomic molecules, each carrying a subset of the genes. Though the size of mtDNA varies between plants, the number of genes appears to be similar in all plants.

In the fully sequenced 367 kilobasepair (kb) mtDNA of *Arabidopsis thaliana,* the known genes (constituting 65 kb) are scattered in a background of noncoding sequence. A similar situation can be found in the sequenced mtDNA of sugarbeet.

Plant mtDNA is quite different from the much smaller (16 kb) single mtDNA molecule of mammals in which every base pair has a function in coding for protein or mediating transcription and replication, and neither introns nor repeats exist. The expression of the mitochondrial genes in plants is carried out by RNA polymerases that are unlike the nuclear polymerases, but related to the enzymes in bacteriophages.

For several mitochondrial genes, these phage-type polymerases start up transcription from multiple promoters. Many genes in plant mtDNA are broken up into exons by introns. The introns are removed in the processing of the primary RNA into a mature mRNA that can be translated into the correct polypeptide.

The introns in mitochondria belong to the so-called self-splicing type II group, in which splicing is defined by the intron structure. The splicing can also work in the *trans* mode, in which it can join segments from separate RNA molecules, and thus assemble protein-coding reading frames even if the exons are distributed on different primary transcripts. This allows individual exons of a gene to reside independently in the genome, even in different subgenomic DNA molecules.

Because mitochondrial genes are often co-transcribed with unrelated genes into a polycistronic primary transcript, RNA processing also includes endonucleolytic reactions, so that each mature mRNA encodes only one protein. Like mRNAs for nuclear encoded genes, mitochondrial mRNAs are polyadenylated. However, instead of stabilizing (the case for nuclear encoded mRNAs), the mitochondrial mRNA polyadenylation mediates degradation. Instead, stem-loops stabilise the mRNA in plant mitochondria.

In plant mitochondria, an additional RNA processing event takes places—RNA editing. In this process, certain cytidines (C) are deaminated into uridines (U). Rarely, U is aminated to C instead.

This means that the coding sequence is changed at the RNA level, in order to restore the correct information for translation where the DNA has diverged. Editing sites mainly occur in DNA encoding protein sequences and functional RNA, and are rare in noncoding mtDNA. In principle, the higher the importance of an individual base, the higher the probability that it constitutes an editing site. Despite the considerable amount of bases that can be edited (up to 15% of the bases of a gene), the functional relevance of this process and the consequent divergence in mtDNA sequence is not understood. Mitochondrial RNA editing has been found in land plants, but not in algae. It also occurs, though to a lesser extent, in plastids.

The mitochondrial genome encodes only a small proportion of the mitochondrial proteins; less than 40 as compared to up to thousands of nuclear-encoded mitochondrial proteins. Through evolution, genes have been transferred from the mitochondrion to the nucleus. This is a process that still takes place, at least in plants.

For example, subunit 2 of cytochrome oxidase is encoded by the mitochondrial genome in all investigated eukaryotes except for some legumes and protists. Within the legume family, several intermediates of the gene transfer have been found. For example, a functional gene can be found in the mitochondrial genome and an unexpressed one in the nuclear genome, or vise versa.

In order for a mitochondrial gene to be successfully transferred to the nucleus, the protein-coding frame must be integrated into the chromosome, and several other modifications must be made.

For example: (1) a promoter for nuclear transcription must be inserted at the beginning of the gene, (2) the mRNA should acquire sequences specifying the addition of a poly-A tail, and (3) a targeting sequence is necessary for the cytosolically synthesized protein to be imported into the mitochondrion. For proper function, elements regulating the expression in different cell types and under different conditions must also be included.

Since the coding sequence for most mitochondrial genes in plants is edited at the RNA level, the mitochondrial genomic sequence will not code for a functional protein if inserted into the nucleus.

Therefore, the DNA that is integrated into the nuclear genome must in most cases be made by reverse transcription from mitochondrial mRNA. Recently, upon sequencing the nuclear genome of *Arabidopsis thaliana*, a copy of most of the mtDNA was found integrated in a nuclear chromosome, indicating that large segments of DNA can move between cellular compartments.

However, most of these mitochondrial genes transferred to the nuclear genome cannot produce functional proteins unless bases corresponding to editing sites are corrected by mutation.

However, this is unlikely to take place before other mutations have further disrupted the sequence. As opposed to other mitochondrial genomes, plant mitochondria utilize the universal genetic code. Thus, this factor will not prevent or delay the transfer of genes between plant organelles as it may do in other eukaryotes.

The special features of the mitochondrial genetic system in plants also have consequences for the study of mitochondrial genes. Due to RNA editing and *trans*-splicing, a reading frame is more difficult to identify based only on genomic sequence.

Also, protein sequences in databases are automatically translated from genomic DNA sequences without compensating for RNA editing in many cases. This protocol would not work for mitochondrial genes.

Does Respiration Reduce Crop Yields?

Plant respiration can consume an appreciable amount of the carbon fixed each day during photosynthesis over and above the losses due to photorespiration. To what extent can changes in a plants respiratory metabolism affect crop yields?

Attempts to establish a quantitative relationship between respiratory energy metabolism and the various processes going on in the cell have led to a break-up of respiration into two components. In growth respiration, reduced carbon is processed to bring about the addition of new biomass. The other component, maintenance respiration, is needed to keep existing, mature cells in a viable state. Utilization of energy by maintenance respiration is not well understood, but estimates indicate that it can represent more than 50% of the total respiratory flux.

Although numerous questions remain regarding these issues, there are several empirical examples of relations between plant respiration rates and crop yield. In the forage crop, perennial ryegrass (*Lolium perenne*), yield increases of 10% to 20% were correlated with a 20% decrease in the leaf respiration rate.

Similar correlations have been found for other plants, including corn (*Zea mays*) and tall fescue (*Festaca arundinacea*). However, later investigations have shown that "selection for low rates of mature leaf respiration is not an appropriate method to select for high-yielding cultivars in perennial grasses".

Although there is a potential for increasing crop yield through reduction of respiration rates, a better understanding of the sites and mechanisms that control plant respiration is needed before such changes can be exploited commercially in a systematic fashion by plant physiologists, geneticists, and molecular biologists.

Furthermore, much remains to be learned about the general applicability of such observations and the conditions under which slower-respiring lines could be at a disadvantage, causing a reduction in crop yields rather than an increase.

Two recent studies illustrate the present difficulty of predicting the effect of directed changes at the molecular level on plant productivity:

Nunes-Nesi et al. (2005) found that the productivity of transgenic tomato plants with a reduced activity of mitochondrial malate dehydrogenase was increased as compared to the wild type. Although the respiratory activity of mitochondria isolated from the transgenic plants was unchanged or higher than in the wild type, the rate of leaf respiration was reduced in the transgenic plants. Photosynthesis was markedly increased in the transgenic lines, possibly linked to increased levels of ascorbate.

Sieger et al. (2005) studied the effect of the alternative oxidase on the growth of tobacco cell cultures. Considering the energy-wasteful nature of the alternative oxidase one might expect that a cell line lacking the enzyme would grow faster than the wild type. Secondly, alternative oxidase has a role in the response of plants to oxidative stress, so a cell line lacking the alternative oxidase would also be expected to handle abiotic stress less well.

The results were surprising—the transgenic cell line with a very low expression of alternative oxidase grew as fast as the wild-type cells under normal, nutrient-sufficient conditions, and faster than the wild type under conditions of macronutrient limitation (low phosphate or low nitrogen).

It appears that the alternative oxidase is an important factor in modulating the growth rate in response to nutrient availability. Plant respiration involves an intricate metabolic network of interacting pathways and a complex regulation of gene expression and enzymatic activities. We clearly need to know much more about these interactions before we can predict the effects of changing the expression of single genes on plant productivity.

The Lipid Composition of Membranes Affects the Cell Biology and Physiology of Plants

The characterization of lipid mutants and the experimental modification of lipid composition by molecular-genetic means have now revealed some broad generalizations about the roles of membrane lipids in the cell biology and physiology of plants. However, the biochemical and physical bases for the phenotypes that are observed remain uncertain. Changes in fatty acid composition have been shown to alter chloroplast size and architecture. Mutations that eliminate polyunsaturated fatty acids block photosynthesis and prevent the plant from growing autotrophically.

Surprisingly, such mutant plants can grow on sucrose, indicating that photosynthesis is the only plant function that absolutely requires a polyunsaturated membrane. However, polyunsaturated fatty acids are also required as precursors of signalling compounds which are necessary for such diverse functions as pollen development and plant defence, such as jasmonic acid,.

One of the most extensively studied issues in membrane biology is the relationship between lipid composition and the ability of organisms to adjust to temperature changes. For example, chill-sensitive plants experience sharp reductions in growth rate and development at temperatures between 0 and 12°C. The physical and physiological changes in chill-sensitive plants that are induced by exposure to low temperature, together with the subsequent expression of stress symptoms, are termed chilling injury.

Many economically important crops, such as cotton, soybean, maize, rice, and many tropical and subtropical fruits, are classified as chill sensitive. In contrast, most plants of temperate origin are able to grow and develop at chilling temperatures and are classified as chill-resistant plants.

In attempts to link the biochemical and physiological changes associated with chilling injury with a single "trigger" or site of damage, it has been suggested that the primary event of chilling injury is a transition from a liquid-crystalline phase to a gel phase in the cellular membranes. According to this proposal, the transition from liquid-crystalline phase to gel phase would result in alterations in the metabolism of chilled cells and lead to injury and death of the chill-sensitive plants.

The degree of unsaturation of the fatty acids would determine the temperature at which such damage occurs. A similar hypothesis has been proposed for chloroplast membranes in which the levels of disaturated phosphatidylglycerol (molecules containing no *cis* double bonds in the fatty acid chains) would determine the chilling sensitivity of plant species.

However, the most recent research indicates that the relationship of membrane unsaturation to plant temperature responses is more subtle and complex than is suggested in these earlier hypotheses. On the one hand, studies of five different *Arabidopsis* mutants have demonstrated that reduced unsaturation can result in plants that grow well at 22°C but are less robust than wild-type plants when grown at 2 to 5°C.

These results were observed even though the lipid changes in most of the mutants are insufficient to cause a lipid phase transition. In addition, the responses of these mutants to low temperature appear quite distinct from classic chilling sensitivity, suggesting that normal chilling injury may not be related to the level of unsaturation of membrane lipids.

In one particular mutant, *fab1*, saturated forms of phosphatidylglycerol account for 43% of the total leaf phosphatidylglycerol—a higher percentage than is found in many chill-sensitive plants. Nevertheless, the mutant was completely unaffected (when compared with wild-type controls) by a range of low-temperature treatments that quickly led to the death of cucumber and other chill-sensitive plants.

A complementary series of experiments has been carried out in tobacco, which is a chill-sensitive plant. The transgenic expression of exogenous genes in tobacco has been used specifically to decrease the level of saturated phosphatidylglycerol or to bring about a general increase in membrane unsaturation.

In each case, damage caused by chilling was alleviated to some extent. These new findings make it clear that the extent of membrane unsaturation or the presence of particular lipids, such as disaturated phosphatidylglycerol, can affect the responses of plants to low temperature. However, membrane lipid composition is not the major determinant of chilling sensitivity in plants.

Utilization of Oil Reserves in Cotyledons

Although the pathway for the mobilization of triacylglycerols has been best characterized in castor bean, it seems to be similar in the storage tissues of other oilseeds.

However, not all seeds quantitatively convert fat to sugar. In castor bean, the endosperm degenerates after the fat and protein reserves are fully utilized. In many oilseeds—such as sunflower (*Helianthus annuus*), cotton (*Gossipium hirsutum*), and members of the squash family (Cucurbitaceae)—the cotyledons become green and photosynthesize after the food reserves are used up. In these tissues only part of the stored lipid is converted to exported carbohydrate.

Much of the lipid-derived carbon remains in the cotyledons, where it contributes to the synthesis of chloroplasts and other cellular structures. Acetyl-CoA is also used directly for the production of energy through respiration. As the greening process takes place, there is a transition in the peroxisome population of these cells: Some peroxisomes have fewer of the characteristics of glyoxysomes and more of leaf-type peroxisomes.

Such a transition is in keeping with the decreased requirement for the breakdown of stored lipids and the increased need to metabolize the products of photorespiration as the tissue goes from a heterotrophic to a more autotrophic mode of metabolism.

Assimilation of Mineral Nutrients

Development of a Root Nodule

The establishment of symbiosis between legumes and rhizobia involves the activation of genes in both the host and the symbiont and an elaborated exchange of signals. The formation of a root nodule, the specialized organ from a plant host that contains the symbiotic nitrogen fixing rhizobia involves two simultaneous processes: infection and nodule organogenesis.

The different stages in the initiation and development of a root nodule are shown in Figure below. A detailed description of the infection process.

A recent study of the early infection stages and nodule development in two *Medicago* species and the symbiont, *Rhisobium meliloti* showed that extensive reorganization of microtubules occurred early in the infection process in the pericycle and the inner cortex where the nodule primordium forms. The changes in microtubule organization were observed by immunolocalization, and fluorescence and laser confocal microscopy.

These findings show that upon infection, the pericycle is activated first, followed by the activation of the inner cortex, where the nodule primordium forms. These changes are followed by a major reorganization of the microtubular cytoskeleton network in outer tissues, associated with root hair activation and curling, the formation of pre-infection threads, and the initiation and the growth of an infection network.

Measurement of Nitrogen Fixation

Besides N_2, the nitrogenase enzyme can reduce other substrates such as cyanide, azide and acetylene.

Direct measurement of N_2 fixation (reduction) requires a mass spectrometer, instrumentation that is not readily available, but the reduction of acetylene to ethylene can easily be measured by gas chromatography. Acetylene reduction involves two electrons, C_2H_2 + $2e^-$ + 2 H^+ → C_2H_4, as opposed to the eight electrons required for the reduction of N_2 and 2 H^+:

$$N_2 + 8e^- + 8H^+ + 16ATP \rightarrow 2NH_3 + H_2 + 16ADP + 16P_i$$

Therefore, reduction of four ethylene molecules correspond to the reduction of one N_2 molecule. The acetylene method has limitations, however: Acetylene blocks gas diffusion into nodules, and rhizosphere microorganisms produce ethylene. For these reasons, the acetylene method is suitable for comparative studies, while mass spectroscopy is used for direct measurements of N_2 reduction, which is required for precise quantification of nitrogen fixation.

The Synthesis of Methionine

Methionine and cysteine are the two sulfur-containing amino acids found in proteins. Methionine is synthesized in plastids from cysteine. In the first step of the pathway, cysteine and O-phosphohomoserine react to form cystathionine via the enzyme cystathionine-γ-synthase. Cystathionine is cleaved into homocysteine, pyruvate, and ammonia by cystathionine-β-lysase. Finally, methionine synthase methylates homocysteine to form methionine (Lea 1997).

Chapter

5

Stress Physiology

Stomatal Conductance and Yields of Irrigated Crops

In the last few decades, the breeding of agricultural crops for higher yields has been very successful. Breeders usually select for high-yielding genotypes empirically, without particular attention to specific plant traits that might be conducive to higher yields. However, comparing old, low-yielding lines of any crop with advanced, high-yielding lines shows clearly that many morphological, physiological, and biochemical traits have been altered by the intense selection pressures for higher yields.

These changes indicate that selection for high-yielding genotypes has generated indirect selection pressures on the altered traits. If we exclude effects of genes that regulate the expression of two or more unrelated traits (pleiotropic genes), the study of high-yielding lines might reveal specific traits and genes associated with the higher yield. With this information, breeders could explicitly select for yield-enhancing traits to improve yields further.

Changes in plant traits associated with yield increases can be studied conveniently in a historical series of a crop. Historical series include successive commercial releases within a breeding program, with each member of the series representing an incremental increase in yield. In studies aimed at characterizing relationships between specific traits and higher yields, all the members of the series are grown together under the same conditions, and the traits of interest, such as plant architecture or photosynthetic rates, are measured in parallel with yields.

Recent studies of this type focusing on historical series of Pima cotton (*Gossypium barbadense*) and bread wheat (*Triticum aestivum*) have shown a remarkable positive correlation between yield increases and increases in stomatal conductance.

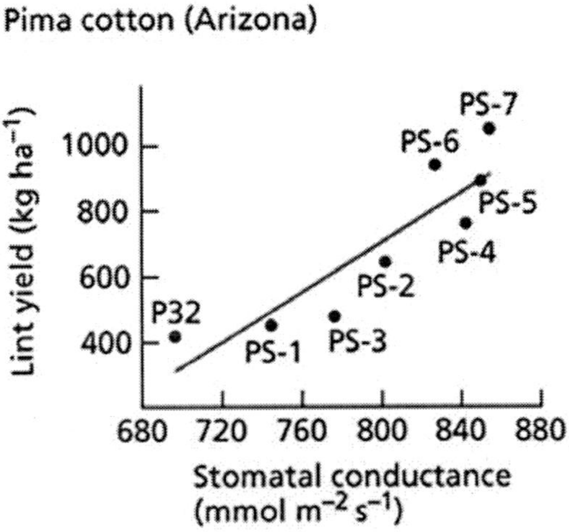

Figure: *Stomatal conductance has increased in parallel with agronomic yields in irrigated Pima cotton (*Gossypium barbadense*) selected for higher yields at a high temperature. The figure shows the relationship between lint yield and stomatal conductance in a historical series of Pima cotton grown in Arizona. The abbreviations P32 and PS-1 through PS-7 designate successive commercial releases between 1949 and 1996.*

Pima cotton is a high-quality fibre cotton grown in hot environments under intensive irrigation. The historical series used in the Pima studies included eight members, released between 1949 and 1996, and encompassing a nearly threefold increase in lint yield. The yield increase attained in each commercial release was accompanied by a corresponding increase in stomatal conductance. Stomatal conductance increased by about 30 mmol m^{-2} s^{-1} for each 100 kg ha^{-1} increase in yield.

Genetic crosses between low-and high-yielding Pima strains have shown that stomatal conductance in Pima cotton has a clear-cut genetic component. The studies with stomata of Pima cotton are the first to show a clear-cut genetic regulation of stomatal conductance proper, of the stomatal response to temperature, and of proton pumping rates of isolated guard cells.

In the absence of explicit selection for higher stomatal conductance in the Pima breeding program, what could be the nature of the indirect selection pressures that caused the increases in conductance paralleling the increases in yield? Higher stomatal conductance increases CO_2 diffusion into the leaf and favours higher photosynthetic rates. Higher photosynthetic rates could in turn favour a higher biomass and higher crop yields. Advanced Pima lines show a higher photosynthetic capacity than older, low-yielding lines, but photosynthetic rates measured in the same leaves used to measure stomatal conductance were not positively correlated with yields. Thus, higher stomatal conductance appears to favour higher yields by a mechanism not directly related to photosynthesis.

Evapotranspiration at the leaf surface lowers leaf temperature, and higher stomatal conductance enhances this leaf cooling. Optimal daytime temperatures for growth, photosynthesis, and reproduction in Pima cotton are below 30°C, while afternoon air temperature in the hot Pima-growing areas often exceeds 40°C. Evaporative cooling of the leaves thus reduces the gap between optimal growth temperatures and air temperature, and provides heat resistance. Studies with the Pima historical series showed that, because of leaf cooling, leaf temperatures were several degrees lower than air temperature, and leaf and canopy temperature were lower in the advanced, high-yielding lines than in low-yielding lines.

The same relationship among yields, stomatal conductance, and leaf temperature was found in a historical series of bread wheat grown in the warm Yaqui Valley of northwestern Mexico. These studies indicate that selection for higher yields in irrigated crops grown at high temperatures imposes indirect selection pressures for high stomatal conductance that lowers leaf temperature and appears to reduce deleterious effects of heat stress on critical flowering and fruiting stages, thus resulting in higher crop yields.

If selection for higher yields generates indirect selection pressures for higher stomatal conductance, can breeders explicitly select for higher conductance and thus obtain lines with higher yields? Studies with Pima cotton have shown that selection of high-conductance F_2 progeny from a cross between high-and low-conductance parents produces high-conductance F_4 lines with higher lint yields than low-conductance lines have. These experiments indicate that high stomatal conductance could be used as a selection trait for high yields in irrigated crops grown at high temperature.

The observed increases in stomatal conductance in crops grown at high temperature also have valuable implications for models developed to predict global climate changes, which are expected to occur because of increases in atmospheric CO_2. Under atmospheric CO_2 concentrations that are twice as high as the present ones, evapotranspiration would decrease over the continents and air temperatures would increase significantly over the tropical land masses, amplifying the changes resulting from atmospheric radiative effects.

However, these models have yet to take into consideration the effect of higher temperatures on the stomatal control of evapotranspiration. Incorporation of the stomatal response to temperature in climate change models is likely to improve the accuracy of the predictions generated by the models. These studies illustrate both the complex interactions between physical and biological factors in the biosphere, and the advantage of a thorough understanding of the physiological properties of plants for efforts aimed at increasing agronomic yields and at improving models that predict global climate changes.

The GABA Shunt Pathway Integrates Stress Metabolic Responses

Not all enzymes participating in metabolic pathways have identical temperature responses, and differential thermostabilities of enzymes may also effect specific steps in metabolism before HSPs acting as molecular chaperones can restore activity. Thus, heat stress can lead to changes in metabolism that result in the accumulation of some metabolites and the reduction of others. Such changes can dramatically alter the function of metabolic pathways and lead to imbalances that are difficult to correct. This problem can be extremely critical when the levels of regulatory metabolites are forced outside of normal ranges so that regulatory homeostasis cannot function properly to restore normal levels.

Additionally, the consumption—and production—of protons during metabolic reactions and/or changes in the activity of membrane-associated proton-transporting ATPases and other membrane-associated proton transporters that move protons out of the cytosol either into the apoplast or vacuoles, can also cause a lowering of the pH of the cytosol, leading to additional metabolic perturbations during stress. Metabolic adaptation to the stress results in the amelioration of such stress effects on metabolism.

One of the metabolic adaptations that plants make to heat stress leads to the accumulation of the nonprotein amino acid, 4-aminobutyric acid (GABA). During episodes of heat stress, GABA accumulates to levels six-to tenfold higher than in unstressed plants. GABA is made from the amino acid, L-glutamate, in a single reaction catalyzed by the enzyme glutamate decarboxylase (GAD). GAD is one of several enzymes whose activity is modulated by the calcium regulatory protein, calmodulin. In the presence of calcium ions, calmodulin is converted into an active state that binds to GAD activating the enzyme, which leads to an increase in the rate at which GABA is produced. The precise mechanism leading to increases in cytosolic calcium have not been directly demonstrated, but studies using inhibitors of calcium ion channels implicate either the plasma membrane or tonoplast calcium channels. Alternatively, the efficiency of one or more of the CAX gene products (calcium proton antiporters) may be reduced as the proton gradient collapses during heat stress, leading to elevated levels of calcium in the cytosol. These increases, in conjunction with calmodulin, activate GAD, producing the high-temperature-induced synthesis of GABA.

GABA is metabolized inside the mitochondria by a GABA aminotransferase (GAT) that utilizes pyruvate and GABA as substrates generating alanine and succinate semialdehyde. This latter compound is then converted into succinate by succinate semialdehyde dehydrogenase (SSA-DH). The function of the mitochondrial components of this pathway leads to the reduction of pyruvate inside the mitochondria and the bypass of citric acid cycle metablism between pyruvate and succinate. This has the effect of bypassing three of the major NADH-producing reactions inside the mitochondria and constitutes a major metabolic adjustment to cellular heat stress that can dramatically alter the redox status of the cell.

Although GABA is an important signalling molcule in mammalian brain tissue, there is no evidence that it directly functions as a signalling molecule in plants. However, the synthesis of GABA consumes protons and thus the GAD-catalyzed reaction is one of several reactions involved in the regulation of cytosolic pH. The function of GAD aids the cell during the heat stress episode by maintaining cytosolic pH in a normal physiological range. Additionally, Arabidopsis mutants of SSA-DH demonstrate increased hydrogen peroxide production during episodes of heat stress and show increased programmed cell death. Thus, the function of the GABA metabolic pathway appears to modulate the oxidative status of the cell during heat stress, and may be involved in the production of a cell signal, effectively modulating PCD.

Membrane Lipids and Low Temperatures

The importance of membrane lipids for tolerance to low temperatures has been demonstrated by work with mutant and transgenic plants in which the activity of particular enzymes led to a specific change in membrane lipid composition independent of acclimation to low temperature.

In one study, the genes used to modify the lipid composition of thylakoid membranes encoded glycerol-3-phosphate acyltransferases (GPATs). These enzymes select fatty acids for incorporation into PG; GPATs from different plant species have different selectivity for saturated and unsaturated fatty acids, leading to the differences in composition.

The degree of fatty acid saturation of phosphatidylglycerol (PG) is particularly important to the chilling resistance of chloroplasts. In one experiment, tobacco was transformed with an acyltransferase gene from either squash (chill sensitive) or *Arabidopsis* (chill tolerant) to modify the fatty acid saturation of PG in the thylakoid membranes of the chloroplasts. In transgenic tobacco transformed with the squash gene, which increased the proportion of high-melting-point (trans-monounsaturated and saturated) molecular species of PG, chilling tolerance decreased (88% inhibition of photosynthesis following 4 hours at 1°C). Conversely, when tobacco was transformed with the *Arabidopsis* chill-tolerant gene, the proportion of high-melting-point molecular species of PG decreased, and tolerance of chilling was greater than in wild-type plants.

Ice Formation in Higher-Plant Cells

During the initial phase of heat loss from tissue, when the temperature drops below the freezing point of the cytosol and the vacuole without ice forming, the liquid phase in the cytosol and vacuole is said to be supercooled. As the temperature drops further, ice forms in the intercellular spaces, and heat energy (0.33 kJ or 80 calories per gram) is released as a result of the latent heat of fusion of water. At this stage, the temperature of the tissue reflects the balance between heat gain from ice formation and heat loss to the environment. As a result, when ice first forms, the temperature rises rapidly (point B in the figure), and it remains at that level until all the extracellular water is frozen (point C in the figure). When that point is reached, heat release stops, and the temperature begins to fall again. The release of heat energy during ice formation (points B to C in the figure) is the basis for the common practice of spraying crops with water during frost: As long as the water continues to freeze extracellularly, it releases heat that prevents intracellular freezing.

The formation of ice within the intercellular spaces of cells that are sensitive to freezing is not lethal, but extended exposure to freezing temperatures causes water vapour to move through the plasma membrane from the unfrozen protoplast to the cell wall, causing ice crystals to grow within the intercellular spaces. This slow dehydration concentrates solutes within the protoplast, depressing the freezing point by 2 to 3°C. As the temperature continues to drop, a second phase of release of the heat of fusion of water is etectable (points from D to E on the figure). This phase reflects a series of small freezing events: Each "spike" of heat release is thought to represent the freezing of cell protoplasts and coincides with loss of viability. The formation of ice crystals on cell walls or in the protoplasm requires the presence of ice nucleation points on which crystals can be initiated and grow.

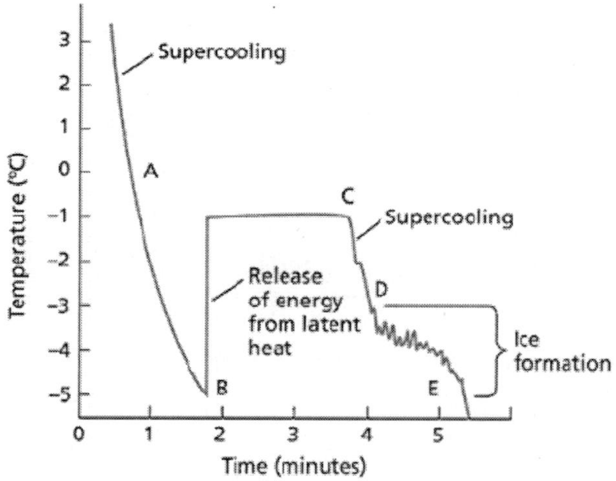

Figure: *Temperature of parenchyma cells in cucumber (*Cucumis sativus*) fruit during freezing. The temperature was recorded with an electronic device, a thermistor, inserted into a 5 × 20 mm cylinder of tissue and immersed in a coolant at –5.8°C. (A–B) Supercooling. (B–C) Release of heat during freezing in cell walls and intercellular spaces. (C–D) Supercooling. (D–E) Small heat spikes released during intracellular freezing of individual protoplasts.*

In some species, acclimation confers an ability to suppress ice nucleation in the protoplast, allowing deep supercooling to many degrees below the freezing point without ice formation. However, deep supercooling of intracellular water has a lower limit of about –40°C, the temperature at which ice forms spontaneously. (The homogeneous ice formation temperature of pure water droplets is –38.1°C, but the presence of solutes lowers this temperature.) At or below –40°C, ice crystals form without nucleation points, and intracellular freezing and cell death are unavoidable. Species that tolerate temperatures below –40°C under natural conditions do so not by supercooling, but by tolerating gradual dehydration.

Plants produce compounds that protect cells against intracellular ice formation. Some of these compounds, such as sugars, amino acids, and other solutes, induce supercooling in the plant's tissues. Overwintering plants also produce antifreeze proteins and other low molecular weight compounds. These compounds provide freeze tolerance by inhibiting ice crystal growth and the nucleation of ice crystals.

Ca^{2+} Signalling and Activation of the Salt Overly Sensitive (SOS) Signal Pathway

Recent studies have identified components of the SOS signal pathway as genetic loci that mediate salt sensitivity. Three genetically linked loci (*SOS1*, *SOS2* and *SOS3*) constitute determinants of a stress-signalling pathway that controls ion homeostasis and salt tolerance.

- SOS3 encodes an EF-hand type Ca^{2+} binding protein with sequence similarity to the regulatory B subunit of calcineurin (protein phosphatase 2B) and neuronal Ca^{2+} sensors. SOS3 has three EF-hand structures, and the interaction of SOS3 with SOS2 kinase leading to activation of SOS2 activity is Ca^{2+} dependent. The function of SOS3 as a salt tolerance determinant *in planta* is dependent on N-myristoylation.
- SOS2 is a serine/threonine protein kinase that is similar to yeast SNF1 and mammalian AMPK protein kinases. Although all the outputs of the SOS signal pathway have yet to be completely defined, genetic and biochemical evidence indicates that the principal effectors regulated by the pathway are transport systems involved in ion homeostasis at least some of which are the same ones discussed above.

- The plasma membrane Na^+/H^+ antiporter, SOS1, is controlled both at the transcriptional and post transcriptional levels by the SOS signalling pathway. Functional disruption of *AtHKT1* was shown to suppress the salt-sensitive phenotype of *sos3*, indicating that the SOS pathway negatively controls this Na^+-influx system. Additionally, the SOS signalling pathway negatively controls expression of vacuolar AtNHX family members.

These studies illustrate the combined use of genetic, physiological, and biochemical approaches for the elucidation of molecular determinants of ion homeostasis and their signal-mediated regulation in plants. Functional genetic studies are investigating the role of these transport proteins in tolerance to salinity stress.

Na^+ Transport across the Plasma Membrane and Vacuolar Compartmentation

The energy-dependent transport of Na^+ from the cytosol of plant cells across the plasma membrane is mediated by a Na^+/H^+ antiporter. This antiporter is known as SOS1 and it is presumably representative of a family of SOS1 genes. *Arabidopsis sos1* mutants are much more sensitive to growth inhibition by salt than are wild type plants. Interestingly, SOS1 gene expression is not regulated by NaCl stress; but it rather depends on other components of the SOS signalling pathway.

Several of the transporters that mediate Na^+ uptake into the cytosol across the plasma membrane have now been identified. In patch clamp studies with isolated membranes, sodium has been shown to compete with K^+ in inward rectifying K^+ channels. Other reports have suggested that K^+ outward-rectifying channels might facilitate Na^+ influx as well. The high affinity K^+ transporter (HKT) from wheat and the low affinity cation transporter (LCT1) from wheat facilitate Na^+ transport in patch clamp studies with isolated membranes, providing evidence to support their possible function in Na^+ uptake. Patch clamp studies have shown the presence of Ca^{2+}-insensitive-sensitive Na^+ conductances in isolated plant plasma membranes.

The Ca^{2+}-sensitive component of Na^+ uptake has been attributed to nonselective cation channels (Schachtman et al. 1997), while *in planta* studies using *Arabidopsis* mutants have directly demonstrated the function of AtHKT1 as a Ca^{2+}-insensitive Na^+-uptake system (Rus et al. 2001). These studies strongly implicate *HKT*1 in the direct control of at least some Na^+ influx. Since *HKT* transcripts are expressed predominantly in the roots, it most probably functions in the control of Na^+ flux into the xylem for export to the shoot.

Vacuolar compartmentalization of Na^+, energized by the ΔpH across the tonoplast membrane, also involves a family of Na^+/H^+ antiporters. *Arabidopsis AtNHX*1 is the prototype for such transporters. Transgenic *Arabidopsis* and tomato plants overexpressing *AtNHX*1 exhibit enhanced salt tolerance. The *Arabidopsis AtNHX*1 family include six loci showing amino acid sequence similarity and topological complementarity to *AtNHX*1, including 12 transmembrane spanning domains and the presence of amiloride binding sites. The *AtNHX*1 family of proteins plays a critical role in the vacuolar compartmentalization of Na^+, which minimizes cytotoxicity and facilitates osmotic adjustment.

Gene Transfer and Stress Tolerance

The transfer of genes suspected to mediate stress tolerance to a test plant where they are overexpressed or altered in some way has given scientists an important new tool with which to study genes involved in stress tolerance. Results of the first study in which transgenic plants were reported to show altered stress tolerance after transfer of a foreign gene appeared in 1993.

The transferred gene encoded the enzyme mannitol dehydrogenase, and its expression led to the accumulation of mannitol in the transgenic plants. Transfer of other genes generated transgenic plants that accumulated other solutes, such as trehalose, proline, and fructan.

Interestingly, the amount of accumulated solutes was insufficient to osmotically balance the applied osmotic stress, yet all of the transgenic plants exhibited some increased stress tolerance. These surprising results are not fully understood, but the newly acquired solutes might protect the transgenic plant against free radicals generated in response to stress, rather than function as osmotic-balancing solutes.

Improved resistance to drought, salinity, and extreme temperatures has been observed in transgenic plants that express/overexpress genes regulating osmolytes, specific proteins, antioxidants, ion homeostasis, transcription factors, and membrane composition.

Plant Physiology

Plant physiology is a subdiscipline of botany concerned with the function, or physiology, of plants. Closely related fields include plant morphology (structure of plants), plant ecology (interactions with the environment), phytochemistry (biochemistry of plants), cell biology, and molecular biology.

Fundamental processes such as photosynthesis, respiration, plant nutrition, plant hormone functions, tropisms, nastic movements, photoperiodism, photomorphogenesis, circadian rhythms, environmental stress physiology, seed germination, dormancy and stomata function and transpiration, both part of plant water relations, are studied by plant physiologists.

Scope

The field of plant physiology includes the study of all the internal activities of plants—those chemical and physical processes associated with life as they occur in plants. This includes study at many levels of scale of size and time. At the smallest scale are molecular interactions of photosynthesis and internal diffusion of water, minerals, and nutrients. At the largest scale are the processes of plant development, seasonality, dormancy, and reproductive control. Major subdisciplines of plant physiology include phytochemistry (the study of the biochemistry of plants) and phytopathology (the study of disease in plants). The scope of plant physiology as a discipline may be divided into several major areas of research.

First, the study of phytochemistry (plant chemistry) is included within the domain of plant physiology. In order to function and survive, plants produce a wide array of chemical compounds not found in other organisms. Photosynthesis requires a large array of pigments, enzymes, and other compounds to function. Because they cannot move, plants must also defend themselves chemically from herbivores, pathogens and competition from other plants. They do this by producing toxins and foul-tasting or smelling chemicals. Other compounds defend plants against disease, permit survival during drought, and prepare plants for dormancy. While other compounds are used to attract pollinators or herbivores to spread ripe seeds.

Secondly, plant physiology includes the study of biological and chemical processes of individual plant cells. Plant cells have a number of features that distinguish them from cells of animals, and which lead to major differences in the way that plant life behaves and responds differently from animal life. For example, plant cells have a cell wall which restricts the shape of plant cells and thereby limits the flexibility and mobility of plants. Plant cells also contain chlorophyll, a chemical compound that interacts with light in a way that enables plants to manufacture their own nutrients rather than consuming other living things as animals do.

Thirdly, plant physiology deals with interactions between cells, tissues, and organs within a plant. Different cells and tissues are physically and chemically specialized to perform different functions. Roots and rhizoids function to anchor the plant and acquire minerals in the soil. Leaves function to catch light in order to manufacture nutrients. For both of these organs to remain living, the minerals acquired by the roots must be transported to the leaves and the nutrients manufactured in the leaves must be transported to the roots. Plants have developed a number of means by which this transport may occur, such as vascular tissue, and the functioning of the various modes of transport is studied by plant physiologists.

Fourthly, plant physiologists study the ways that plants control or regulate internal functions. Like animals, plants produce chemicals called hormones which are produced in one part of the plant to signal cells in another part of the plant to respond. Many flowering plants bloom at the appropriate time because of light-sensitive compounds that respond to the length of the night, a phenomenon known as photoperiodism. The ripening of fruit and loss of leaves in the winter are controlled in part by the production of the gas ethylene by the plant.

Finally, plant physiology includes the study of how plants respond to conditions and variation in the environment, a field known as environmental physiology. Stress from water loss, changes in air chemistry, or crowding by other plants can lead to changes in the way a plant functions. These changes may be affected by genetic, chemical, and physical factors.

Biochemistry of Plants

The list of simple elements of which plants are primarily constructed—carbon, oxygen, hydrogen, calcium, phosphorus, etc.—is not different from similar lists for animals, fungi, or even bacteria. The fundamental atomic components of plants are the same as for all life; only the details of the way in which they are assembled differs.

Despite this underlying similarity, plants produce a vast array of chemical compounds with unusual properties which they use to cope with their environment. Pigments are used by plants to absorb or detect light, and are extracted by humans for use in dyes. Other plant products may be used for the manufacture of commercially important rubber or biofuel. Perhaps the most celebrated compounds from plants are those with pharmacological activity, such as salicylic acid (aspirin), morphine, and digitalis. Drug companies spend billions of dollars each year researching plant compounds for potential medicinal benefits.

Constituent Elements

Plants require some nutrients, such as carbon and nitrogen, in large quantities to survive. Such nutrients are termed macronutrients, where the prefix *macro*-(large) refers to the quantity needed, not the size of the nutrient particles themselves. Other nutrients, called micronutrients, are required only in trace amounts for plants to remain healthy. Such micronutrients are usually absorbed as ions dissolved in water taken from the soil, though carnivorous plants acquire some of their micronutrients from captured prey.

The following tables list element nutrients essential to plants. Uses within plants are generalized.

Table: *Macronutrients. (Necessary in large quantities)*

Element	***Form of uptake***	***Notes***
Nitrogen	NO_3^-, NH_4^+	Nucleic acids, proteins, hormones, etc.
Oxygen	O_2 H_2O	Cellulose, starch, other organic compounds
Carbon	CO_2	Cellulose, starch, other organic compounds
Hydrogen	H_2O	Cellulose, starch, other organic compounds
Potassium	K^+	Cofactor in protein synthesis, water balance, etc.
Calcium	Ca^{2+}	Membrane synthesis and stabilization
Magnesium	Mg^{2+}	Element essential for chlorophyll
Phosphorus	$H_2PO_4^-$	Nucleic acids, phospholipids, ATP
Sulfur	SO_4^{2-}	Constituent of proteins and coenzymes

Table: *Micronutrients. (Necessary in small quantities)*

Element	***Form of uptake***	***Notes***
Chlorine	Cl^-	Photosystem II and stomata function
Iron	Fe^{2+}, Fe^{3+}	Chorophyll formation
Boron	HBO_3	Crosslinking pectin

Manganese	Mn^{2+}	Activity of some enzymes
Zinc	Zn^{2+}	Involved in the synthesis of enzymes and chlorophyll
Copper	Cu^{+}	Enzymes for lignin synthesis
Molybdenum	MoO_4^{2-}	Nitrogen fixation, reduction of nitrates
Nickel	Ni^{2+}	Enzymatic cofactor in the metabolism of nitrogen compounds

Pigments

Among the most important molecules for plant function are the pigments. Plant pigments include a variety of different kinds of molecules, including porphyrins, carotenoids, and anthocyanins. All biological pigments selectively absorb certain wavelengths of light while reflecting others. The light that is absorbed may be used by the plant to power chemical reactions, while the reflected wavelengths of light determine the colour the pigment will appear to the eye.

Chlorophyll is the primary pigment in plants; it is a porphyrin that absorbs red and blue wavelengths of light while reflecting green. It is the presence and relative abundance of chlorophyll that gives plants their green colour. All land plants and green algae possess two forms of this pigment: chlorophyll *a* and chlorophyll *b*. Kelps, diatoms, and other photosynthetic heterokonts contain chlorophyll *c* instead of *b*, while red algae possess only chlorophyll *a*. All chlorophylls serve as the primary means plants use to intercept light in order to fuel photosynthesis.

Carotenoids are red, orange, or yellow tetraterpenoids. They function as accessory pigments in plants, helping to fuel photosynthesis by gathering wavelengths of light not readily absorbed by chlorophyll. The most familiar carotenoids are carotene (an orange pigment found in carrots), lutein (a yellow pigment found in fruits and vegetables), and lycopene (the red pigment responsible for the colour of tomatoes). Carotenoids have been shown to act as antioxidants and to promote healthy eyesight in humans.

Anthocyanins (literally "flower blue") are water-soluble flavonoid pigments that appear red to blue, according to pH. They occur in all tissues of higher plants, providing colour in leaves, stems, roots, flowers, and fruits, though not always in sufficient quantities to be noticeable. Anthocyanins are most visible in the petals of flowers, where they may make up as much as 30% of the dry weight of the tissue. They are also responsible for the purple colour seen on the underside of tropical shade plants such as *Tradescantia zebrina*; in these plants, the anthocyanin catches light that has passed through the leaf and reflects it back towards regions bearing chlorophyll, in order to maximize the use of available light.

Betalains are red or yellow pigments. Like anthocyanins they are water-soluble, but unlike anthocyanins they are indole-derived compounds synthesized from tyrosine. This class of pigments is found only in the Caryophyllales (including cactus and amaranth), and never co-occur in plants with anthocyanins. Betalains are responsible for the deep red colour of beets, and are used commercially as food-colouring agents. Plant physiologists are uncertain of the function that betalains have in plants which possess them, but there is some preliminary evidence that they may have fungicidal properties.

Signals and Regulators

Plants produce hormones and other growth regulators which act to signal a physiological response in their tissues. They also produce compounds such as phytochrome that are sensitive to light and which serve to trigger growth or development in response to environmental signals.

Plant Hormones

Plant hormones, also known as plant growth regulators (PGRs) or phytohormones, are chemicals that regulate a plant's growth. According to a standard animal definition, hormones are signal molecules produced at specific locations, that occur in very low concentrations, and cause altered processes in target cells at other locations. Unlike animals, plants lack specific hormone-producing tissues or organs. Plant hormones are often not transported to other parts of the plant and production is not limited to specific locations.

Plant hormones are chemicals that in small amounts promote and influence the growth, development and differentiation of cells and tissues. Hormones are vital to plant growth; affecting processes in plants from flowering to seed development, dormancy, and germination. They regulate which tissues grow upwards and which grow downwards, leaf formation and stem growth, fruit development and ripening, as well as leaf abscission and even plant death.

The most important plant hormones are abscissic acid (ABA), auxins, gibberellins, and cytokinins, though there are many other substances that serve to regulate plant physiology.

Photomorphogenesis

While most people know that light is important for photosynthesis in plants, few realize that plant sensitivity to light plays a role in the control of plant structural development (morphogenesis). The use of light to control structural development is called photomorphogenesis, and is dependent upon the presence of specialized photoreceptors, which are chemical pigments capable of absorbing specific wavelengths of light.

Plants use four kinds of photoreceptors: phytochrome, cryptochrome, a UV-B photoreceptor, and protochlorophyllide *a*. The first two of these, phytochrome and cryptochrome, are photoreceptor proteins, complex molecular structures formed by joining a protein with a light-sensitive pigment. Cryptochrome is also known as the UV-A photoreceptor, because it absorbs ultraviolet light in the long wave "A" region.

The UV-B receptor is one or more compounds that have yet to be identified with certainty, though some evidence suggests carotene or riboflavin as candidates. Protochlorophyllide *a*, as its name suggests, is a chemical precursor of chlorophyll.

The most studied of the photoreceptors in plants is phytochrome. It is sensitive to light in the red and far-red region of the visible spectrum. Many flowering plants use it to regulate the time of flowering based on the length of day and night (photoperiodism) and to set circadian rhythms. It also regulates other responses including the germination of seeds, elongation of seedlings, the size, shape and number of leaves, the synthesis of chlorophyll, and the straightening of the epicotyl or hypocotyl hook of dicot seedlings.

Photoperiodism

Many flowering plants use the pigment phytochrome to sense seasonal changes in day length, which they take as signals to flower. This sensitivity to day length is termed photoperiodism. Broadly speaking, flowering plants can be classified as long day plants, short day plants, or day neutral plants, depending on their particular response to changes in day length. Long day plants require a certain minimum length of daylight to initiate flowering, so these plants flower in the spring or summer. Conversely, short day plants will flower when the length of daylight falls below a certain critical level. Day neutral plants do not initiate flowering based on photoperiodism, though some may use temperature sensitivity (vernalization) instead.

Although a short day plant cannot flower during the long days of summer, it is not actually the period of light exposure that limits flowering. Rather, a short day plant requires a minimal length of uninterrupted darkness in each 24 hour period (a short daylength) before floral development can begin. It has been determined experimentally that a short day plant (long night) will not flower if a flash of phytochrome activating light is used on the plant during the night.

Plants make use of the phytochrome system to sense day length or photoperiod. This fact is utilized by florists and greenhouse gardeners to control and even induce flowering out of season, such as the *Poinsettia*.

Environmental Physiology

Paradoxically, the subdiscipline of environmental physiology is on the one hand a recent field of study in plant ecology and on the other hand one of the oldest. Environmental physiology is the preferred name of the subdiscipline among plant physiologists, but it goes by a number of other names in the applied sciences. It is roughly synonymous with ecophysiology, crop ecology, horticulture and agronomy. The particular name applied to the subdiscipline is specific to the viewpoint and goals of research. Whatever name is applied, it deals with the ways in which plants respond to their environment and so overlaps with the field of ecology.

Environmental physiologists examine plant response to physical factors such as radiation (including light and ultraviolet radiation), temperature, fire, and wind. Of particular importance are water relations (which can be measured with the Pressure bomb) and the stress of drought or inundation, exchange of gases with the atmosphere, as well as the cycing of nutrients such as nitrogen and carbon.

Environmental physiologists also examine plant response to biological factors. This includes not only negative interactions, such as competition, herbivory, disease and parasitism, but also positive interactions, such as mutualism and pollination.

Tropisms and Nastic Movements

Plants may respond both to directional and nondirectional stimuli. A response to a directional stimulus, such as gravity or sunlight, is called a tropism. A response to a nondirectional stimulus, such as temperature or humidity, is a nastic movement.

Tropisms in plants are the result of differential cell growth, in which the cells on one side of the plant elongate more than those on the other side, causing the part to bend toward the side with less growth. Among the common tropisms seen in plants is phototropism, the bending of the plant toward a source of light. Phototropism allows the plant to maximize light exposure in plants which require additional light for photosynthesis, or to minimize it in plants subjected to intense light and heat. Geotropism allows the roots of a plant to determine the direction of gravity and grow downwards. Tropisms generally result from an interaction between the environment and production of one or more plant hormones.

In contrast to tropisms, nastic movements result from changes in turgor pressure within plant tissues, and may occur rapidly. A familiar example is thigmonasty (response to touch) in the Venus fly trap, a carnivorous plant. The traps consist of modified leaf blades which bear sensitive trigger hairs. When the hairs are touched by an insect or other animal, the leaf folds shut. This mechanism allows the plant to trap and digest small insects for additional nutrients. Although the trap is rapidly shut by changes in internal cell pressures, the leaf must grow slowly in order to reset for a second opportunity to trap insects.

Plant Disease

Economically, one of the most important areas of research in environmental physiology is that of phytopathology, the study of diseases in plants and the manner in which plants resist or cope with infection. Plant are susceptible to the same kinds of disease organisms as animals, including viruses, bacteria, and fungi, as well as physical invasion by insects and roundworms. Because the biology of plants differs from animals, their symptoms and responses are quite different. In some cases, a plant can simply shed infected leaves or flowers to prevent to spread of disease, in a process called abscission.

Most animals do not have this option as a means of controlling disease. Plant diseases organisms themselves also differ from those causing disease in animals because plants cannot usually spread infection through casual physical contact. Plant pathogens tend to spread via spores or are carried by animal vectors.

One of the most important advances in the control of plant disease was the discovery of Bordeaux mixture in the nineteenth century. The mixture is the first known fungicide and is a combination of copper sulfate and lime. Application of the mixture served to inhibit the growth of downy mildew that threatened to seriously damage the French wine industry.

History

Early History

Sir Francis Bacon published one of the first plant physiology experiments in 1627 in the book, *Sylva Sylvarum.* Bacon grew several terrestrial plants, including a rose, in water and concluded that soil was only needed to keep the plant upright. Jan Baptist van Helmont published what is considered the first quantitative experiment in plant physiology in 1648. He grew a willow tree for five years in a pot containing 200 pounds of oven-dry soil.

The soil lost just two ounces of dry weight and van Helmont concluded that plants get all their weight from water, not soil. In 1699, John Woodward published experiments on growth of spearmint in different sources of water. He found that plants grew much better in water with soil added than in distilled water. Stephen Hales is considered the Father of Plant Physiology for the many experiments in the 1727 book; though Julius von Sachs unified the pieces of plant physiology and put them together as a discipline. His *Lehrbuch der Botanik* was the plant physiology bible of its time.

Researchers discovered in the 1800s that plants absorb essential mineral nutrients as inorganic ions in water. In natural conditions, soil acts as a mineral nutrient reservoir but the soil itself is not essential to plant growth. When the mineral nutrients in the soil are dissolved in water, plant roots absorb nutrients readily, soil is no longer required for the plant to thrive. This observation is the basis for hydroponics, the growing of plants in a water solution rather than soil, which has become a standard technique in biological research, teaching lab exercises, crop production and as a hobby.

Current Research

One of the leading journals in the field is *Plant Physiology*, started in 1926. All its back issues are available online for free. Many other journals often carry plant physiology articles, including *Physiologia Plantarum*, *Journal of Experimental Botany*, *American Journal of Botany*, *Annals of Botany*, *Journal of Plant Nutrition* and *Proceedings of the National Academy of Sciences.*

Economic Applications

Food Production

In horticulture and agriculture along with food science, plant physiology is an important topic relating to fruits, vegetables, and other consumable parts of plants. Topics studied include: *climatic* requirements, fruit drop, nutrition, ripening, fruit set. The production of food crops also hinges on the study of plant physiology covering such topics as Optimal planting and harvesting times and post harvest storage of plant products for human consumption and the production of secondary products like drugs and cosmetics.

Plant Nutrition

Plant nutrition is the study of the chemical elements that are necessary for plant growth. There are several principles that apply to plant nutrition. Some elements are directly involved in plant metabolism. However, this principle does not account for the so-called beneficial elements, whose presence, while not required, has clear positive effects on plant growth.

A nutrient that is able to limit plant growth according to Liebig's law of the minimum, is considered an essential plant nutrient if the plant can not complete its full life cycle without it. There are 14 essential plant nutrients.

Carbon, Hydrogen and Oxygen are essential for plants, but are not considered plant nutrients.

Macronutrients:

- N = Nitrogen
- K = Potassium
- Ca = Calcium
- P = Phosphorus
- Mg = Magnesium
- S = Sulfur
- Si = Silicon

Micronutrients (trace levels) include:

- Cl = Chlorine
- Fe = Iron
- B = Boron
- Mn = Manganese

- Na = Sodium
- Zn = Zinc
- Cu = Copper
- Ni= Nickel
- Mo = Molybdenum.

These nutrients are further divided into the mobile and immobile nutrients. A plant will always supply more nutrients to its younger leaves than its older ones, so when nutrients are mobile, the lack of nutrients is first visible on older leaves. When a nutrient is less mobile, the younger leaves suffer because the nutrient does not move up to them but stays lower in the older leaves. Nitrogen, phosphorus, and potassium are mobile nutrients, while the others have varying degrees of mobility.

Processes

Plants uptake essential elements from the soil through their roots and from the air through their leaves. Nutrient uptake in the soil is achieved by cation exchange, wherein root hairs pump hydrogen ions (H^+) into the soil through proton pumps. These hydrogen ions displace cations attached to negatively charged soil particles so that the cations are available for uptake by the root. In the leaves, stomata open to take in carbon dioxide and expel oxygen. The carbon dioxide molecules are used as the carbon source in photosynthesis.

Though nitrogen is plentiful in the earth's atmosphere, relatively few plants engage in nitrogen fixation (conversion of atmospheric nitrogen to a biologically useful form). Most plants therefore require nitrogen compounds to be present in the soil in which they grow.

Functions of Nutrients

Each of these nutrients is used in a different place for a different essential function.

Carbon: Carbon forms the backbone of many plants biomolecules, including starches and cellulose. Carbon is fixed through photosynthesis from the carbon dioxide in the air and is a part of the carbohydrates that store energy in the plant.

Hydrogen: Hydrogen also is necessary for building sugars and building the plant. It is obtained almost entirely from water.

Oxygen: Oxygen is necessary for cellular respiration. Cellular respiration is the process of generating energy-rich adenosine triphosphate (ATP) via the consumption of sugars made in photosynthesis. Oxygen gas is produced as a by-product from this reaction.

Phosphorus: Phosphorus is important in plant bioenergetics. As a component of ATP, phosphorus is needed for the conversion of light energy to chemical energy (ATP) during photosynthesis. Phosphorus can also be used to modify the activity of various enzymes by phosphorylation, and can be used for cell signalling.

Since ATP can be used for the biosynthesis of many plant biomolecules, phosphorus is important for plant growth and flower/ seed formation.

Potassium: Potassium regulates the opening and closing of the stoma by a potassium ion pump. Since stomata are important in water regulation, potassium reduces water loss from the leaves and increases drought tolerance. Potassium deficiency may cause necrosis or interveinal chlorosis.

Nitrogen: Nitrogen is an essential component of all proteins. Nitrogen defiiency most often results in stunted growth.

Sulphur: Sulphur is a structural component of some amino acids and vitamins, and is essential in the manufacturing of chloroplasts.

Silicon: Silicon is deposited in cell walls and contributes to its mechanical properties including rigidity and elasticity

Calcium: Calcium regulates transport of other nutrients into the plant and is also involved in the activation of certain plant enzymes. Calcium deficiency results in stunting.

Magnesium: Magnesium is an important part of chlorophyll, a critical plant pigment important in photosynthesis. It is important in the production of ATP through its role as an enzyme cofactor. There are many other biological roles for magnesium. Magnesium deficiency can result in interveinal chlorosis.

Iron: Iron is necessary for photosynthesis and is present as an enzyme cofactor in plants. Iron deficiency can result in interveinal chlorosis and necrosis.

Molybdenum: Molybdenum is a cofactor to enzymes important in building amino acids.

Boron: Boron is important in sugar transport, cell division, and synthesizing certain enzymes. Boron deficiency causes necrosis in young leaves and stunting.

Copper: Copper is important for photosynthesis. Symptoms for copper deficiency include chlorosis. Involved in many enzyme processes. Necessary for proper photosythesis. Involved in the manufacture of lignin (cell walls). Involved in grain production.

Manganese: Manganese is necessary for building the chloroplasts. Manganese deficiency may result in coloration abnormalities, such as discoloured spots on the foliage.

Sodium: Sodium is involved in the regeneration of phosphoenol-pyruvate in CAM and C4 plants. It can also substitute for potassium in some circumstances.

Zinc: Zinc is required in a large number of enzymes and plays an essential role in DNA transcription. A typical symptom of zinc deficiency is the stunted growth of leaves, commonly known as "little leaf" and is caused by the oxidative degradation of the growth hormone auxin.

Nickel: In higher plants, Nickel is essential for activation of urease, an enzyme involved with nitrogen metabolism that is required to process urea. Without Nickel, toxic levels of urea accumulate, leading to the formation of necrotic lesions. In lower plants, Nickel activates several enzymes involved in a variety of processes, and can substitute for Zinc and Iron as a cofactor in some enzymes.

Cobalt has proven to be beneficial to at least some plants, but is essential in others, such as legumes where it is required for nitrogen fixation. Vanadium may be required by some plants, but at very low concentrations. It may also be substituting for molybdenum. Selenium and sodium may also be beneficial. Sodium can replace potassium's regulation of stomatal opening and closing.

Plant nutrition is a difficult subject to understand completely, partially because of the variation between different plants and even between different species or individuals of a given clone. Elements present at low levels may cause deficiency symptoms, and toxicity is possible at levels that are too high. Further, deficiency of one element may present as symptoms of toxicity from another element, and vice-versa. Carbon and oxygen are absorbed from the air, while other nutrients are absorbed from the soil. Green plants obtain their carbohydrate supply from the carbon dioxide in the air by the process of photosynthesis.

Plant Hormone

Plant hormones (also known as phytohormones) are chemicals that regulate plant growth. In the UK, the term 'hormone' is not acceptable.

There they are called 'plant growth substances'. Plant hormones are signal molecules produced within the plant, and occur in extremely low concentrations. Hormones regulate cellular processes in targeted cells locally and when moved to other locations, in other locations of the plant. Hormones also determine the formation of flowers, stems, leaves, the shedding of leaves, and the development and ripening of fruit. Plants, unlike animals, lack glands that produce and secrete hormones. Plant hormones shape the plant, affecting seed growth, time of flowering, the sex of flowers, senescence of leaves and fruits. They affect which tissues grow upward and which grow downward, leaf formation and stem growth, fruit development and ripening, plant longevity and even plant death. Hormones are vital to plant growth and lacking them, plants would be mostly a mass of undifferentiated cells.

Characteristics

The word hormone is derived from Greek and means 'set in motion.' Plant hormones affect gene expression and transcription levels, cellular division and growth. They are naturally produced within plants, though very similar chemicals are produced by fungi and bacteria that can also effect plant growth. A large number of related chemical compounds are synthesized by humans, they are used to regulate the growth of cultivated plants, weeds, and in vitro grown plants and plant cells; these man made compounds are called Plant Growth Regulators or PGRs for short. Early in the study of plant hormones, "phytohormone" was the commonly-used term, but its use is less widely applied now.

Plant hormones are not nutrients, but chemicals that in small amounts promote and influence the growth, development, and differentiation of cells and tissues. The biosynthesis of plant hormones within plant tissues is often diffuse and not always localized. Plants lack glands to produce and store hormones, because, unlike animals, which have two circulatory systems (lymphatic and cardiovascular) powered by a heart that moves fluids around the body, plants use more passive means to move chemicals around the plant. Plants utilize simple chemicals as hormones, which move more easily through the plant's tissues. They are often produced and used on a local basis within the plant body, plant cells even produce hormones that affect different regions of the cell producing the hormone.

Hormones are transported within the plant by utilizing four types of movements. For localized movement, cytoplasmic streaming within cells and slow diffusion of ions and molecules between cells are utilized. Vascular tissues are used to move hormones from one part of the plant to another; these include sieve tubes that move sugars from the leaves to the roots and flowers, and xylem that moves water and mineral solutes from the roots to the foliage.

Not all plant cells respond to hormones, but those cells that do are programmed to respond at specific points in their growth cycle. The greatest effects occur at specific stages during the cell's life, with diminished effects occurring before or after this period. Plants need hormones at very specific times during plant growth and at specific locations. They also need to disengage the effects that hormones have when they are no longer needed. The production of hormones occurs very often at sites of active growth within the meristems, before cells have fully differentiated. After production they are sometimes moved to other parts of the plant where they cause an immediate effect or they can be stored in cells to be released later. Plants use different pathways to regulate internal hormone quantities and moderate their effects; they can regulate the amount of chemicals used to biosynthesize hormones. They can store them in cells, inactivate them, or cannibalise already-formed hormones by conjugating them with carbohydrates, amino acids or peptides. Plants can also break down hormones chemically, effectively destroying them. Plants also move hormones around the plant diluting their concentrations.

The concentration of hormones required for plant responses are very low (10^{-6} to 10^{-5} mol/L). Because of these low concentrations it has been very difficult to study plant hormones and only since the late 1970s have scientists been able to start piecing together their effects and relationships to plant physiology. Much of the early work on plant hormones involved studying plants that were genetically deficient in one or involved the use of tissue cultured plants grown *in vitro* that were subjected to differing ratios of hormones and the resultant growth compared. The earliest scientific observation and study dates to the 1880s; the determination and observation of plant hormones and their identification was spread-out over the next 70 years.

Classes of Plant Hormones

It is generally accepted that there are five major classes of plant hormones, some of which are made up of many different chemicals that can vary in structure from one plant to the next. The chemicals are each grouped together into one of these classes based on their structural similarities and on their effects on plant physiology. Other plant hormones and growth regulators are not easily grouped into these classes, they exist naturally or are synthesized by humans or other organisms, including chemicals that inhibit plant growth or interrupt the physiological processes within plants. Each class has positive as well as inhibitory functions, and most often work in tandem with each other, with varying ratios of one or more interplaying to affect growth regulation.

The five major classes are:

Abscisic Acid

Abscisic acid also called ABA, was discovered and researched under two different names before its chemical properties were fully known, it was called *dormin* and *abscicin II.* Once it was determined that the two latter named compounds were the same, it was named abscisic acid. The name "abscisic acid" was given because it was found in high concentrations in newly-abscissed or freshly-fallen leaves.

This class of PGR is composed of one chemical compound normally produced in the leaves of plants, originating from chloroplasts, especially when plants are under stress. In general, it acts as an inhibitory chemical compound that affects bud growth, seed and bud dormancy. It mediates changes within the apical meristem causing bud dormancy and the alteration of the last set of leaves into protective bud covers. Since it was found in freshly-abscissed leaves, it was thought to play a role in the processes of natural leaf drop but further research has disproven this. In plant species from temperate parts of the world it plays a role in leaf and seed dormancy by inhibiting growth, but, as it is dissipated from seeds or buds, growth begins. In other plants, as ABA levels decrease, growth then commences as gibberellin levels increase. Without ABA, buds and seeds would start to grow during warm periods in winter and be killed when it froze again. Since ABA dissipates slowly from the tissues and its effects take time to be offset by other plant hormones, there is a delay in physiological pathways that provide some protection from premature growth. It accumulates within seeds during fruit maturation, preventing seed germination within the fruit, or seed germination before winter. Abscisic acid's effects are degraded within plant tissues during cold temperatures or by its removal by water washing in out of the tissues, releasing the seeds and buds from dormancy.

In plants under water stress ABA plays a role in closing the stomata. Soon after plants are water stressed and the roots are deficient in water, a signal moves up to the leaves causing the formation of ABA precursors there which then move to the roots. The roots then release ABA which is translocated to the foliage through the vascular system and modulates the potassium and sodium uptake within the guard cells, which then lose turgidity, closing the stomata. ABA exists in all parts of the plant and its concentration within any tissue seems to mediate its effects and function as a hormone, its degradation or more properly catabolism within the plant affects metabolic reactions and cellular growth and production of other hormones.

Plants start life as a seed with high ABA levels, just before the seed germinates ABA levels decrease; during germination and early growth of the seedling, ABA levels decrease even more. As plants begin to produce shoots with fully functional leaves-ABA levels begin to increase, slowing down cellular growth in more "mature" areas of the plant. Stress from water or predation affects ABA production and catabolism rates which mediate another cascade of effects triggering specific responses from targeted cells. Scientists are still piecing together the complex interactions and effects of this and other phytohormones.

Auxins

Auxins are compounds that positively influence cell enlargement, bud formation and root initiation. They also promote the production of other hormones and in conjunction with cytokinins, they control the growth of stems, roots, fruits and convert stems into flowers. Auxins were the first class of growth regulators discovered. They affect cell elongation by altering cell wall plasticity. Auxins decrease in light and increase where its dark. They stimulate cambium cells to divide and in stems cause secondary xylem to differentiate. Auxins act to inhibit the growth of buds lower down the stems (apical dominance), and also to promote lateral and adventitious root development and growth. Leaf abscission is initiated by the growing point of a plant ceasing to produce auxins. Auxins in seeds regulate specific protein synthesis, as they develop within the flower after pollination, causing the flower to develop a fruit to contain the developing seeds. Auxins are toxic to plants in large concentrations; they are most toxic to dicots and less so to monocots. Because of this property, synthetic auxin herbicides including 2,4-D and 2,4,5-T have been developed and used for weed control. Auxins, especially 1-

Naphthaleneacetic acid (NAA) and Indole-3-butyric acid (IBA), are also commonly applied to stimulate root growth when taking cuttings of plants. The most common auxin found in plants is indoleacetic acid or IAA.

Cytokinins

Cytokinins or CKs are a group of chemicals that influence cell division and shoot formation. They were called kinins in the past when the first cytokinins were isolated from yeast cells. They also help delay senescence or the aging of tissues, are responsible for mediating auxin transport throughout the plant, and affect internodal length and leaf growth. They have a highly-synergistic effect in concert with auxins and the ratios of these two groups of plant hormones affect most major growth periods during a plant's lifetime. Cytokinins counter the apical dominance induced by auxins; they in conjunction with ethylene promote abscission of leaves, flower parts and fruits.

Ethylene

Ethylene is a gas that forms through the Yang Cycle from the breakdown of methionine, which is in all cells. Ethylene has very limited solubility in water and does not accumulate within the cell but diffuses out of the cell and escapes out of the plant. Its effectiveness as a plant hormone is dependent on its rate of production versus its rate of escaping into the atmosphere. Ethylene is produced at a faster rate in rapidly growing and dividing cells, especially in darkness. New growth and newly-germinated seedlings produce more ethylene than can escape the plant, which leads to elevated amounts of ethylene, inhibiting leaf expansion. As the new shoot is exposed to light, reactions by phytochrome in the plant's cells produce a signal for ethylene production to decrease, allowing leaf expansion. Ethylene affects cell growth and cell shape; when a growing shoot hits an obstacle while underground, ethylene production greatly increases, preventing cell elongation and causing the stem to swell. The resulting thicker stem can exert more pressure against the object impeding its path to the surface. If the shoot does not reach the surface and the ethylene stimulus becomes prolonged, it affects the stems natural geotropic response, which is to grow upright, allowing it to grow around an object. Studies seem to indicate that ethylene affects stem diameter and height: When stems of trees are subjected to wind, causing lateral stress, greater ethylene production occurs, resulting in thicker, more sturdy tree trunks and branches. Ethylene affects fruit-ripening:

Normally, when the seeds are mature, ethylene production increases and builds-up within the fruit, resulting in a climacteric event just before seed dispersal. The nuclear protein ETHYLENE INSENSITIVE2 (EIN2) is regulated by ethylene production, and, in turn, regulates other hormones including ABA and stress hormones.

Gibberellins

Gibberellins or GAs include a large range of chemicals that are produced naturally within plants and by fungi. They were first discovered when Japanese researchers, including Eiichi Kurosawa, noticed a chemical produced by a fungus called *Gibberella fujikuroi* that produced abnormal growth in rice plants. Gibberellins are important in seed germination, affecting enzyme production which mobilizes food production used for growth of new cells.

This is done by modulating chromosomal transcription. In grain (rice, wheat, corn, etc.) seeds, a layer of cells called the aleurone layer wraps around the endosperm tissue. Absoption of water by the seed causes production of GA. The GA is transported to the aleurone layer, which responds by producing enzymes that break down stored food reserves within the endosperm, which are utilized by the growing seedling. GAs produce bolting of rosette-forming plants, increasing internodal length. They promote flowering, cellular division, and in seeds growth after germination. Gibberellins also reverse the inhibition of shoot growth and dormancy induced by ABA.

Other Known Hormones

Other identified plant growth regulators include:

- Brassinolides-plant steroids that are chemically similar to animal steroid hormones. First isolated from pollen of the mustard family and extensively studied in *Arabidopsis*. They promote cell elongation and cell division, differentiation of xylem tissues, and inhibit leaf abscission. Plants that are deficient in brassinolides suffer from dwarfism.
- Salicylic acid-activates genes in some plants that produce chemicals that aid in the defence against pathogenic invaders.
- Jasmonates-are produced from fatty acids and seem to promote the production of defence proteins that are used to fend off invading organisms. They are believed to also have a role in seed germination, and affect the storage of protein in seeds, and seem to affect root growth.

- Plant peptide hormones-encompass all small secreted peptides that are involved in cell-to-cell signalling. These small peptide hormones play crucial roles in plant growth and development, including defence mechanisms, the control of cell division and expansion, and pollen self-incompatibility.
- Polyamines-are strongly basic molecules with low molecular weight that have been found in all organisms studied thus far. They are essential for plant growth and development and affect the process of mitosis and meiosis.
- Nitric oxide (NO)-serves as signal in hormonal and defence responses.
- Strigolactones, implicated in the inhibition of shoot branching.

Potential Medical Applications

Plant stress hormones activate cellular responses, including cell death, to diverse stress situations in plants. Researchers have found that some plant stress hormones share the ability to adversely affect human cancer cells.

For example, sodium salicylate has been found to suppress proliferation of lymphoblastic leukemia, prostate, breast, and melanoma human cancer cells. Jasmonic acid, a plant stress hormone that belongs to the jasmonate family, induced death in lymphoblastic leukemia cells.

Methyl jasmonate has been found to induce cell death in a number of cancer cell lines.

Hormones and Plant Propagation

Synthetic plant hormones or PGRs are commonly used in a number of different techniques involving plant propagation from cuttings, grafting, micropropagation, and tissue culture.

The propagation of plants by cuttings of fully-developed leaves, stems, or roots is performed by gardeners utilizing auxin as a rooting compound applied to the cut surface; the auxins are taken into the plant and promote root initiation. In grafting, auxin promotes callus tissue formation, which joins the surfaces of the graft together. In micropropagation, different PGRs are used to promote multiplication and then rooting of new plantlets. In the tissue-culturing of plant cells, PGRs are used to produce callus growth, multiplication, and rooting.

Seed Dormancy

Plant hormones affect seed germination and dormancy by affecting different parts of the seed.

Embryo dormancy is characterized by a high ABA/GA ratio, whereas the seed has a high ABA sensitivity and low GA sensitivity. To release the seed from this type of dormancy and initiate seed germination, an alteration in hormone biosynthesis and degradation towards a low ABA/GA ratio, along with a decrease in ABA sensitivity and an increase in GA sensitivity needs to occur.

ABA controls embryo dormancy, and GA embryo germination. Seed coat dormancy involves the mechanical restriction of the seed coat, this along with a low embryo growth potential, effectively produces seed dormancy. GA releases this dormancy by increasing the embryo growth potential, and/or weakening the seed coat so the radical of the seedling can break through the seed coat. Different types of seed coats can be made up of living or dead cells and both types can be influenced by hormones; those composed of living cells are acted upon after seed formation while the seed coats composed of dead cells can be influenced by hormones during the formation of the seed coat. ABA affects testa or seed coat growth characteristics, including thickness, and effects the GA-mediated embryo growth potential.

These conditions and effects occur during the formation of the seed, often in response to environmental conditions. Hormones also mediate endosperm dormancy: Endosperm in most seeds is composed of living tissue that can actively respond to hormones generated by the embryo. The endosperm often acts as a barrier to seed germination, playing a part in seed coat dormancy or in the germination process. Living cells respond to and also affect the ABA/GA ratio, and mediate cellular sensitivity; GA thus increases the embryo growth potential and can promote endosperm weakening. GA also affects both ABA-independent and ABA-inhibiting processes within the endosperm.

History of Plant Physiology

Plant physiology is concerned with the life processes of plants, and from the beginning has been focused largely on the higher green terrestrial plants, the autotrophic (self-feeding) plants that feed us animals. In part, plant physiology has roots in agriculture.

The central question of plant physiology is how do plants grow, develop, and reproduce? When primitive humans collected seeds and began raising food plants they must have noted that plants need sunlight, warmth and moist (but not wet) soil of good tilth, and that seeds from vigorous plants produced vigorous plants. They observed the beneficial effects of manuring (mentioned in the Bible; Luke 13:8). Centuries of agricultural practice produced improved varieties and cultural practices, and early studies of physiology drew on this basic knowledge of plant growth and gross anatomy.

Early Experiments on Growth

An early physiological question was where a plant gets the material with which it grows. In the early 1600s, Jan van Helmont, a Belgian physician, decided the source must be water alone. Van Helmont grew a willow seedling in 200 pounds of soil, and only added rainwater. A 164-pound tree was produced with only 57.1 grams (2 ounces) of soil lost. He knew of carbon dioxide but never dreamed that a diffuse gas could produce willow wood.

In the next century Antoine Lavoisier found organic matter to be largely formed of carbon and oxygen. Joseph Priestley, Jan Ingenhousz, and Jean Senebier demonstrated that plant leaves in light take up carbon dioxide and emit equivalent amounts of oxygen. Later, Nicholas de Saussure noted that water was involved in the process. The reverse occurred in the dark—plants respired like animals, taking up oxygen and emitting carbon dioxide. J. R. Mayer observed that the process converted light energy into the chemical energy of organic carbon. Thus growth of seedlings in the dark or of roots in the soil was at the expense of this energy. Therefore, by the nineteenth century, photosynthesis, although not understood biochemically, was established as the primary and essential synthetic process in plant growth.

Nutrition and Transportation

In his experiment, van Helmont assigned no importance to the two ounces of soil lost. However, starting in the late 1700s and extending into the mid 1800s, Julius Sachs and others used chemical assays to establish that quantitatively minor soil constituents of nitrogen, potassium, phosphate, sulfur, and other elements had major importance in plant growth.

The long-recognized importance of manure lay with its content of these inorganic nutrients, especially nitrogen. It was discovered these could be added to the soil as inorganic salts, such as potassium nitrate. The organic material of manure, or the residue of its decay, contributed to improved tilth, or soil structure, but did not provide nutrients. From these discoveries came the modern agricultural use of chemical fertilizers.

What about the extensive loss of water from the soil? Van Helmont had to continuously water his willow tree with many more pounds of water than were ultimately incorporated by the tree. In 1727 an English clergyman and amateur physiologist, Stephen Hales, published *Vegetable Staticks,* an account of his pioneering studies on the transpiration, growth, and gas exchanges of plants. Hales demonstrated that water from the soil moves up the stems to the leaves where it is lost as water vapour, a process called transpiration. Subsequent research of the nineteenth and early twentieth centuries showed that the water diffuses out through stomata (singular stoma), pores in the leaf epidermis (outer layer of leaf cells).

With light and adequate water the two cells bounding the stoma inflate, opening the pore to gas diffusion; under dry conditions the cells grow flaccid and the pore closes, conserving water. Capillary forces originating in the microscopic pores of the leaf mesophyll (internal green photosynthetic cells), with some contribution from osmosis, pull columns of water up the open vessels and tracheids of the xylem (wood) carrying nutrient salts from the roots. The coherence between water molecules and their adherence to cell walls prevents the taut water columns from breaking even in trees of great height. This scheme was first proposed in 1895 by Henry Dixen and John Joly. Numerous researchers in the twentieth century confirmed and refined this "cohesion-tension" theory of transportation.

Hales also measured the root pressure (forced bleeding) of decapitated plants. Subsequent work showed that under conditions of good soil moisture and aeration, roots actively secrete high concentrations of salt into the root xylem creating a high osmotic pressure that forces water up the stem and out pores at the tips of leaves (guttation). In 1926, E. Munch proposed a similar mechanism for translocation, the movement of sugars from leaves to roots and other plant parts. This mechanism is known as the pressureflow model.

Cellular and Molecular Plant Physiology

By the twentieth century, plant physiologists increasingly turned to chemistry and physics for assistance with fundamental questions. They also established their own societies with journals to publish their findings, which had a catalytic effect in increasing the level and amount of research. A great deal of the fundamental biochemistry of cell growth and function, known from the more extensive medical, animal, and microbiological research, was found to apply to plant cells. Anatomical studies gave structural details to support physiological findings, and submicroscopic cell structure was revealed by the electron microscope.

All the mineral nutrients required for plant growth were established. The key to their selective uptake from the soil and transport into the root xylem proved to lie with an energy-requiring proton (hydrogen ion) pumping mechanism in the cell membranes.

Environmental, hormonal, and genetic controls on growth and development have been extensively explored, but there is still more to learn. Ethylene, a simple two-carbon gas generated by plants initiates fruit ripening and regulates aspects of seed germination. Phototropisms (bending in response to unilateral light, investigated by Charles Darwin), and geotropisms (root growth down, stem growth up) were found to be due to displacement of a cell growth hormone, or auxin. In some circumstances, auxins could also elicit cell division (root formation in stem cuttings). Other hormones, the gibberellins, regulate cell division at the stem apex and activate enzyme formation in seed germination.

Attempts to culture plant tissues led to discovery of more cell division hormones, the cytokinins. Another type of hormone, abscisic acid, initiates the senescence and abscission of leaves in the fall, and causes the stomata to close under water stress. Additional growth regulating compounds are being found and investigated but a coordinated picture of hormone interaction is lacking.

Photoperiodism, the regulation of flowering by day length was discovered. Sleep movements, such as the drooping of bean leaves in the evening, were found to be controlled by a biological "clock," a circadian rhythm, not by the onset of darkness. In 1952, phytochrome was discovered and found to be the pigment at the centre of photoperiodism.

In recent years there has been a major shift to molecular genetics in attempts to locate the genes responsible for physiological processes. In photosynthesis chlorophyll structure was determined and localized in the internal membranes of the chloroplasts of the mesophyll cells. Red and blue portions of the light spectrum were found effective, leading to the discovery that two light reactions are required. In the 1930s, C. B. van Neil used radioactive water to show that water, not carbon dioxide, was the source of oxygen released during photosynthesis. Sugar was found to be synthesized in the stroma (fluid part) of the chloroplast, and the molecular details of its creation were worked out by Melvin Calvin and Andrew Benson. All plant cells were found to respire, an energy-yielding process essentially the same as that in animals, involving another membranous organelle, the mitochondrion, and yielding metabolic energy available for transport reactions and synthesis of cell substance.

The formation of fats and oils from carbohydrates was found to be similar to that in animals, but plants had the added ability to transform oils in germinating seeds into carbohydrates such as the glucose used in cellulose wall formation. The symbiotic relationships of plants and microorganisms was explored, notably in the cases of reduced nitrogen formation from atmospheric nitrogen by nodule bacteria.

At the end of twentieth century, the small mustard plant *Arabidopsis thaliana* took centre stage in the attempt of scientists to understand plant genomes. The full sequence of this genome was elucidated in 2000 by an international consortium of plant geneticists.

Plant Cell

Plant cells are eukaryotic cells that differ in several key respects from the cells of other eukaryotic organisms. Their distinctive features include:

- A large central vacuole, a water-filled volume enclosed by a membrane known as the *tonoplast* maintains the cell's turgor, controls movement of molecules between the cytosol and sap, stores useful material and digests waste proteins and organelles.
- A cell wall composed of cellulose and hemicellulose, pectin and in many cases lignin, and secreted by the protoplast on the outside of the cell membrane. This contrasts with the cell walls of fungi (which are made of chitin), and of bacteria, which are made of peptidoglycan.
- Specialised cell-cell communication pathways known as plasmodesmata, pores in the primary cell wall through which the plasmalemma and endoplasmic reticulum of adjacent cells are continuous.

- Plastids, notably the chloroplasts which contain chlorophyll and the biochemical systems for light harvesting and photosynthesis, but also amyloplasts specialized for starch storage, elaioplasts specialized for fat storage and chromoplasts specialized for synthesis and storage of pigments. As in mitochondria, which have a genome encoding 37 genes plastids have their own genomes of about 100-120 unique genes and probably arose as prokaryotic endosymbionts living in the cells of an early eukaryotic ancestor of the land plants and algae.
- Cell division by construction of a phragmoplast as a template for building a cell plate late in cytokinesis is characteristic of land plants and a few groups of algae, notably the Charophytes and the Order Trentepohliales
- The sperm of bryophytes have flagellae similar to those in animals, but higher plants, (including Gymnosperms and flowering plants) lack the flagellae and centrioles that are present in animal cells.

Cell Types

- Parenchyma cells are living cells that have diverse functions ranging from storage and support to photosynthesis and phloem loading (transfer cells). Apart from the xylem and phloem in its vascular bundles, leaves are mainly composed of parenchyma cells. Some parenchyma cells, as in the epidermis, are specialized for light penetration and focusing or regulation of gas exchange, but others are among the least specialized cells in plant tissue, and may remain totipotent, capable of dividing to produce new populations of undifferentiated cells, throughout their lives. Parenchyma cells have thin, permeable primary walls enabling the transport of small molecules between them, and their cytoplasm is responsible for a wide range of biochemical functions such as nectar secretion, or the manufacture of secondary products that discourage herbivory. Parenchyma cells which contain many chloroplasts and are concerned primarily with photosynthesis are called chlorenchyma cells. Others, such as the majority of the parenchyma cells in potato tubers and the seed cotyledons of legumes have a storage function.
- Collenchyma cells-Like parenchyma, collenchyma cells are alive at maturity and have only a primary wall. These cells mature

from meristem derivatives that initially resemble parenchyma, but differences quickly become apparent. Plastids do not develop and the secretory apparatus (ER and Golgi) proliferates to secrete additional primary wall. The wall is most commonly thickest at the corners, where three or more cells come in contact and thinnest where only two cells come in contact, though other arrangements of the wall thickening are possible.

Pectin and hemicellulose are the dominant constituents of collenchyma cell walls of dicotyledon angiosperms which may contain as little as 20% of cellulose in *Petasites*. Collenchyma cells are typically quite elongated, and may divide transversly to give a septate appearance. The role of this cell type is to support the plant in axes still growing in length, and to confer flexibility and tensile strength on tissues. The primary wall lacks lignin that would make it tough and rigid, so this cell type provides what could be called plastic support. Support that can hold a young stem or petiole into the air, but in cells that can be stretched as the cells around them elongate. Stretchable support (without elastic snap-back) is a good way to describe what collenchyma does. Parts of the strings in celery are collenchyma.

- Sclerenchyma cells-Sclerenchyma cells (from the Greek skleros, hard) are hard and tough cells with a function in mechanical support. They are of two broad types-sclereids or stone cells and fibres. The cells develop an extensive secondary cell wall that is laid down on the inside of the primary cell wall. The secondary wall is impregnated with lignin, making it hard and impermeable to water. Thus, these cells cannot survive for long as they cannot exchange sufficient material to maintain active metabolism. Sclerenchyma cells are typically dead at functional maturity, and the cytoplasm is missing, leaving an empty central cavity.

Functions for sclereid cells (hard cells that give leaves or fruits a gritty texture) include discouraging herbivory, by damaging digestive passages in small insect larval stages, and physical protection (a solid tissue of hard sclereid cells form the pit wall in a peach and many other fruits). Functions of fibres include provision of load-bearing support and tensile strength to the leaves and stems of herbaceous plants. Sclerenchyma fibres are not involved in conduction, either of water and nutrients (as in the xylem) or of carbon compounds (as in the phloem), but it is likely that they may have evolved as modifications of xylem and phloem initials in early land plants.

Tissue Types

The major classes of cells differentiate from undifferentiated meristematic cells (analogous to the stem cells of animals) to form the tissue structures of roots, stems, leaves, flowers and reproductive structures.

Xylem cells are elongated cells with lignified secondary thickening of the cell walls. Xylem cells are specialised for conduction of water, and first appeared in plants during their transition to land in the Silurian period more than 425 million years ago. The possession of xylem defines the vascular plants or Tracheophytes. Xylem tracheids are pointed, elongated xylem cells, the simplest of which have continuous primary cell walls and lignified secondary wall thickenings in the form of rings, hoops or reticulate networks.

More complex tracheids with valve-like perforations called bordered pits characterise the gymnosperms. The ferns and other pteridophytes and the gymnosperms only have xylem tracheids, while the angiosperms also have xylem vessels. Vessel members are hollow xylem cells aligned end-to-end, without end walls that are assembled into long continuous tubes. The bryophytes lack true xylem cells, but their sporophytes have a water conducting tissue known as the hydrome that is composed of elongated cells of simpler construction.

Phloem is a specialised tissue for food conduction in higher plants. Phloem consists of two cell types, the sieve tubes and the intimately-associated companion cells. The sieve tube elements lack nuclei and ribosomes, and their metabolism and functions are regulated by the adjacent nucleate companion cells. Sieve tubes are joined end to end with perforate end-plates between known as *sieve plates*, which allow transport of photosynthate between the sieve elements. The companion cells, connected to the sieve tubes via plasmodesmata, are responsible for loading the phloem with sugars. The bryophytes lack phloem, but moss sporophytes have a simpler tissue with analogous function known as the leptome.

Plant epidermal cells are specialised parenchyma cells covering the external surfaces of leaves, stems and roots. The epidermal cells of aerial organs arise from the superficial layer of cells known as the *tunica* that covers the plant shoot apex, whereas the cortex and vascular tissues arise from innermost layer of the shoot apex known as the *corpus* (L3 layer). The epidermis of roots originates from the layer of cells immediately beneath the root cap.

The epidermis of all aerial organs, but not roots, is covered with a cuticle made of waxes and the polyester cutin. Several cell types may be present in the epidermis. Notable among these are the stomatal guard cells, glandular and clothing hairs or trichomes, and the root hairs of primary roots. In the shoot epidermis of most plants, only the guard cells have chloroplasts. The epidermal cells of the primary shoot are thought to be the only plant cells with the biochemical capacity to synthesize cutin.

The Plant Cell

Plant cells are eukaryotic and have many of the structures found in animal cells. Link to pages describing these.

- Plasma membrane
- Nucleus and nucleolus
- Mitochondria
- Ribosomes
- Endoplasmic reticulum
- Golgi apparatus
- Peroxisomes (the crystal in the electron micrograph is enclosed within a peroxisome)
- Microtubules.

Plant cells differ from animal cells in lacking:

- centrioles
- intermediate filaments.

and having:

- plastids
- a cell wall
- large vacuoles.

The electron micrograph shows cells from a sunflower leaf. It was supplied through the courtesy of H. J. Arnott and Kenneth M. Smith.

Vacuoles

Vacuoles are bounded by a single membrane. Young plant cells often contain many small vacuoles, but as the cells mature, these unite to form a large central vacuole. Vacuoles serve several functions, such as

- storing foods (e.g., proteins in seeds)
- storing wastes

- storing malic acid in CAM plants
- storing various ions (e.g., calcium, sodium, iron) which, among other functions, helps to
- maintain turgor in the cell.

Plant cells avoid bursting in hypotonic surroundings by their strong cell walls. These allow the build-up of turgor within the cell. Loss of turgor causes wilting.

Plasmolysis

When a freshwater (or terrestrial) plant is placed in sea water, its cells quickly lose turgor and the plant wilts.

This is because sea water is hypertonic to the cytoplasm. As water diffuses from the cytoplasm into the sea water, the cells shrink-drawing their plasma membrane away from the cell wall.

The photomicrograph shows plasmolyzed cells in the freshwater plant Elodea which has been placed in sea water. Note how the cell walls now show clearly.

Chapter

6

Phytochrome and Light Control of Plant Development

The Structure of Phytochromes

Methods for phytochrome purification were developed through research pioneered by the agricultural biochemist H. William Siegelman, working at the USDA lab at Beltsville, Maryland. After surveying many different plants and plant tissues, Siegelman found that the young etiolated seedlings of cereals, such as oat and rye, were particularly rich sources of phytochrome.

Some initial success in purifying phytochrome from etiolated oats (*Avena sativa*) was reported in 1964, and many workers subsequently used dark-grown oat seedlings for phytochrome extractions. As a result, much of the biochemical information on phytochrome is derived from studies on oat seedlings, although there is a considerable amount of data on phytochrome isolated from dark-grown rye (*Secale* spp.) and pea (*Pisum sativum*) seedlings as well.

While on a fresh-weight basis there is 50 times more phytochrome in etiolated tissue than in equivalent green tissue of the same age, the overall content of phytochrome in etiolated plants is low—about 0.2% of the total extractable protein. Thus, to purify even milligram quantities usually requires more than a kilogram fresh weight of starting material. Etiolated seedlings are also rich in proteases, and much of the phytochrome is susceptible to proteolysis. To minimize the loss of phytochrome, researchers carry out these extractions rapidly, at a low temperature, and in the presence of protease inhibitors. The stability of phytochrome during purification is also favoured by conversion to the Pfr form with red light. For these reasons, successful purification of fully intact, native phytochrome from etiolated oat seedlings was not reported until 1983.

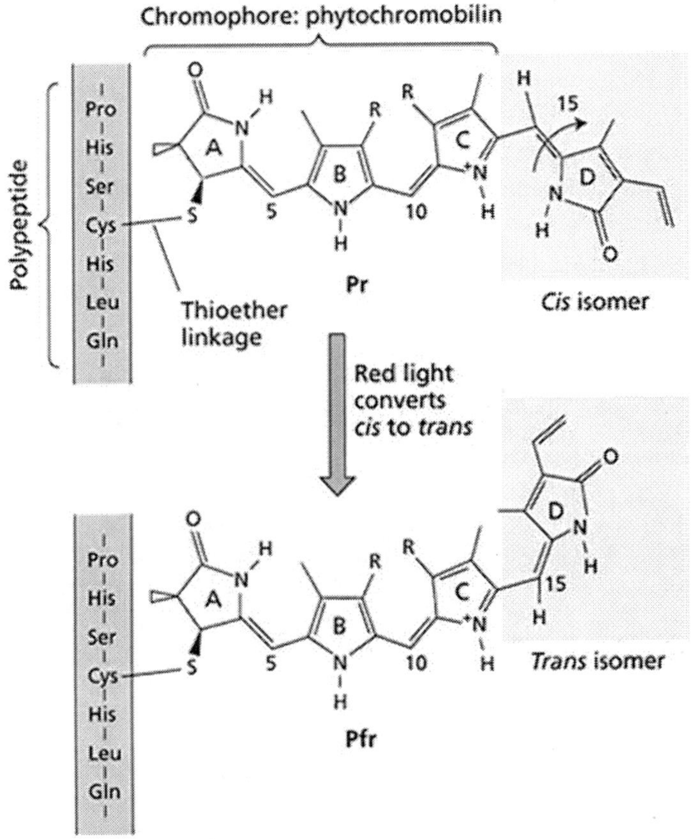

Figure: *Structure of the Pr and Pfr forms of the chromophore (phytochromobilin) and the peptide region bound to the chromophore through a thioether linkage. The chromophore undergoes a* cis-trans *isomerization at carbon 15 in response to red and far-red light. (After Andel et al. 1997.)*

Native phytochrome is a soluble protein with a molecular mass of about 250 kDa. It occurs as a dimer made up of two equivalent subunits. Each subunit consists of two components: a light-absorbing pigment molecule called the chromophore, and a polypeptide chain called the apoprotein. The apoprotein monomer has a molecular mass of about 125 kDa. Together, the apoprotein and its chromophore make up the holoprotein. The chromophore of phytochrome is a linear tetrapyrrole termed phytochromobilin. There is only one chromophore per monomer of apoprotein, and it is attached to the protein through a thioether linkage to a cysteine residue.

Researchers have visualized the Pr form of phytochrome using electron microscopy by spraying a solution of the protein onto mica flakes, drying it under vacuum, and "shadowing" the dried molecules by using platinum atoms to enhance its three-dimensional structure. As Figure.B shows, phytochrome molecules appear as a collection of spots, the number depending on the orientation of the protein. On the basis of this information and more accurate measurements using X-ray scattering, the model has been proposed. The polypeptide folds into two major domains separated by a "hinge" region. The larger N-terminal domain is approximately 70 kDa and bears the chromophore; the smaller C-terminal domain is approximately 55 kDa and contains the site where the two monomers associate with each other to form the dimer.

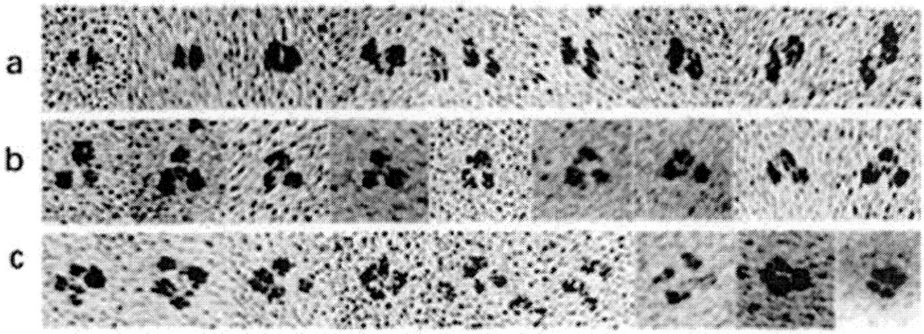

Figure: *Phytochrome from pea, purified and then sprayed onto a mica flake. After being dried, the specimen was shadowed with platinum to enhance the three-dimensional structure. A carbon replica was prepared and viewed in the electron microscope. Various images were produced, including pairs of spots (a), three spots (b), and clusters of spots (c). The variation in the pattern in A is caused by different orientations of the protein. (After Tokutomi et al. 1989, courtesy of M. Wada.)*

Phytochrome and High-Irradiance Responses

HIRs, such as the inhibition of stem or hypocotyl growth, have usually been studied in dark-grown, etiolated seedlings. The HIR action spectrum for the inihibition of hypocotyl elongation in dark-grown lettuce seedlings. Note that the main peak of activity is in the far-red rather than the red region of the spectrum, and there are peaks in the blue and UV-A regions as well. Because the absence of a peak in the red is unusual for a phytochome-mediated response, at first researchers believed that another pigment might be involved. In the late 1960s, Karl M. Hartmann in Freiburg, Germany proposed a model to explain the apparent discrepancy. Taking a cue from the enhancement effect of photosynthesis that was discovered by R. Emerson and coworkers, Hartmann speculated that the HIR might be the result of an interaction between the Pr and Pfr forms of phytochrome.

To probe such an interaction, Hartmann irradiated plants with not one but two wavelengths of light simultaneously, using two wavelengths that were inactive in the HIR for lettuce hypocotyl inhibition: 658 nm and 768 nm. Although physiologically inactive, the two wavelengths are still *photochemically* active in driving phytochrome conversions, each in the opposite direction. By carefully varying the irradiances of each wavelength, Hartmann was able to induce a range of ratios of Pfr to total phytochrome in the plant from near zero to higher values. He was able to show that the optimum ratio of Pfr to total phytochrome for the inhibition of lettuce hypocotyl growth is approximately 0.03. Remarkably, this is precisely the ratio that results when the plants are irradiated with a single wavelength at 720 nm, the far-red peak of the HIR.

The Origins of Phytochrome as a Bacterial Two-Component Receptor

Bacteria often use two-component signalling systems to sense and respond to environmental signals such as nutrients or light. The first component is a sensor protein, consisting of an *input domain* that receives the input signal, linked to a *transmitter domain* that transmits the signal to the second component. The second component is the response regulator protein, composed of a *receiver domain* that accepts the signal from the transmitter of the sensor protein, linked to an *output domain* that initiates the cellular response.

These two components interact by a series of phosphorylation and dephosphorylation reactions. When it receives a signal, the sensor autophosphorylates a specific histidine residue in the transmitter domain. This phosphate is then transferred to a specific aspartate residue on the receiver domain of the response regulator. But what has all this to do with phytochrome?

Although they lack the specific histidine residue that becomes autophosphorylated, the C termini of phytochromes have some sequence similarity to the transmitter domains of bacterial sensor proteins. This sharing of sequences raises the interesting possibility that phytochrome evolved from the sensor protein of a bacterial two-component system.

Provocative new evidence for the idea that phytochrome evolved from the histidine kinase of a bacterial two-component system has recently emerged from studies of cyanobacteria. The cyanobacterium *Fremyella* uses light to regulate the composition of its photosynthetic pigments contained within the light-harvesting phycobilisomes, a response termed chromatic adaptation.

Like phytochromes, phycobiliproteins bind linear tetrapyrrole chromophores that are closely related in structure to the phytochrome chromophore, but there is no evidence for red/far-red reversible responses of the type seen in higher plants. Instead, chromatic adaptation occurs in response to growth in red light and green light.

When the gene for the photoreceptor for this response (*RcaE*) was cloned, it was found to encode a 74 kDa polypeptide, the N-terminal end of which is related to the N terminus of higher-plant phytochromes, while the C-terminal end is related to histidine kinases. However, the lack of the conserved chromophore-binding site and of red/far-red reversibility suggests that this protein is a distant relative of higher-plant phytochromes.

Publication of the complete sequence of the genome of the cyanobacterium *Synechocystis* in 1996 led to great excitement because this organism contains a gene that has between 50 and 60 amino acid identity with *Arabidopsis* phyE along its length, including a bona fide chromophore attachment site. In add tion, the amino acid sequence of the C terminus of the *Synechocystis* pigment protein is related to the KinA sensor protein of *Bacillus subtilis*. This *Synechocystis* gene has been named *cph1* (cyanobacterial *ph*ytochrome *1*).

Recently it was shown that the purified Cph1 apoprotein can assemble with phycocyanobilin, a cyanobacterial linear tetrapyrrole, in vitro, and the resultant holoprotein exhibits red/far-red reversibility and absorption spectra similar (but not identical) to higher-plant phytochrome assembled in vitro with the same chromophore.

Perhaps the most exciting finding of all was the identification, 10 base pairs downstream from the cyanobacterial phytochrome gene, of another gene having the features of a bacterial response regulator. Since the C-terminal end of the cyanobacterial phytochrome contains all the conserved features of a histidine kinase transmitter module, and the downstream response regulator resembles aspartate kinase receiver modules, it was now possible to test whether this phytochrome acts as part of a light-regulated transmitter–receiver pair, as had been proposed earlier.

Using the purified proteins, the Cph1 protein was shown to phosphorylate itself in a manner consistent with its having a histidine kinase activity, and further, this phosphate could be transferred to the response regulator. Surprisingly, Cph1 is far more active as a kinase in the Pr form than in the Pfr form, suggesting that the light signal is transduced through regulation of Pr abundance, rather than through Pfr abundance as in higher plants.

Taken together, the results from these studies suggest that plant phytochromes originally evolved from a class of photoreversible transmitter histidine kinases in cyanobacteria. Are plant phytochromes also kinases? In support of this idea, a phytochrome gene isolated from the moss *Ceratodon* contains an additional C-terminal sequence that has kinase homology. In addition, highly purified preparations of higher-plant phytochrome have been found to contain a polycation-stimulated protein kinase activity that is also greater in the Pr form than in the Pfr form.

However, the kinase activity associated with these phytochromes is of the serine–threonine type, and consistent with this observation, these proteins lack the conserved histidines of the bacterial sensor proteins. More recently, it was shown that recombinant oat phyA is a serine-threonine kinase with the same biochemical properties as the plant-derived photoreceptor.

Moreover, R-light stimulated its autophosphorylation, supporting the hypothesis that Pfr is the active form. Polycations were shown to stimulate Pr phosphorylation more than Pfr, a result consistent with previous studies.

Thus, while higher-plant phytochromes share a homology with bacterial two-component signalling systems, the reaction mechanism and components may have diverged considerably. Nevertheless, the discovery of cyanobacterial phytochromes has opened up many exciting new areas of research that have already contributed much to our understanding of phytochrome evolution and function.

Mougeotia: A Chloroplast with a Twist

Mougeotia is a filamentous green alga whose cells contain a single ribbonlike chloroplast nestled between two large vacuoles and surrounded by a layer of cytoplasm. The chloroplast can rotate about its long axis and can respond to light by orienting *perpendicularly* to the direction of light. The perpendicular orientation is referred to as "face position."

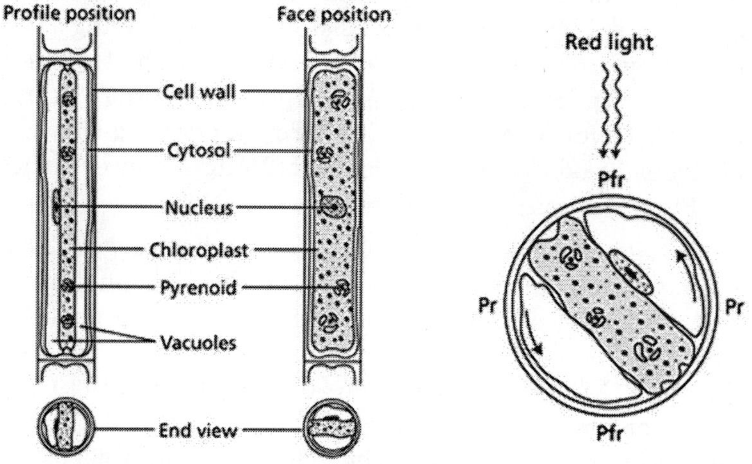

Phytochrome is implicated in the *Mougeotia* chloroplast rotation response because the system is most sensitive to red light and because the red light effect can be reversed by far-red light. Using microbeams of red and far-red light, Wolfgang Haupt and his coworkers in Germany showed that in *Mougeotia*, phytochrome is localized near the periphery of the cytoplasm in the vicinity of the plasma membrane. When a microbeam of red light was directed at the cell surface under a microscope, the edge of the chloroplast adjacent to the microbeam rotated 90°, even though the chloroplast itself was not illuminated.

Experiments with Plane Polarized Light

The orientation of individual phytochrome molecules within the cell was probed by the use of plane-polarized light. Each phytochrome molecule has a preferred orientation for the absorption of plane-polarized light. The greatest absorption occurs when this light is parallel to the absorbing surface of the pigment.

If we think of the pigment as a rod and the polarized light as a narrow slit of light, more light will be absorbed when the light slit is parallel to the long axis of the pigment than in any other orientation.

Haupt and his coworkers discovered that the ability of plane-polarized red light to cause chloroplast rotation depends on the plane of polarization relative to the long axis of the cell. This phenomenon is depicted in the experiment shown in below. Half of a cell was illuminated over the entire surface with polarized red light (R) vibrating *perpendicular* to the cell axis; the other half was illuminated with light vibrating *parallel* to the cell axis.

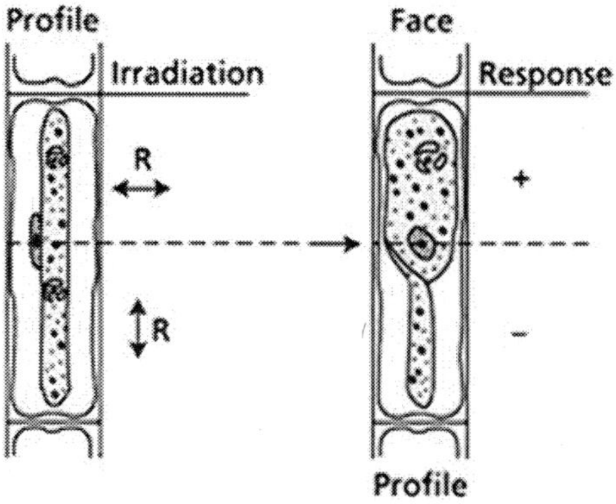

The chloroplast rotated from the profile to the face position only in response to red light vibrating in the perpendicular direction, producing a twist in the chloroplast. On the basis of these experiments, it was concluded that phytochrome molecules have a defined orientation. Two additional experiments demonstrated that the phytochrome orientation changes during the conversion from Pr to Pfr. In the first experiment, shown in Figure below, the cytoplasmic layer was irradiated with a microbeam of plane-polarized red light vibrating either *parallel* (upper light treatment) or *perpendicular* (lower light treatment) to the cell's long axis. In apparent contradiction to the results shown in Figure above, the chloroplast rotated only when the red light was vibrating *parallel* to the cell's long axis.

The apparent contradiction is due to the cylindrical shape of the cell. Thus, the phytochrome molecules near the two edges will be oriented 90° relative to the phytochrome molecules at the centre. The phytochrome molecules that absorb red light when the polarized microbeam is aimed at the edge of the cell in the experiments of Figure above, will be oriented 90° from the phytochrome molecules that absorb light in the *whole-cell* irradiation experiments. In the second experiment, shown on the right-hand side of Figure above, the cytoplasmic layer on the right side of the cell was irradiated with a microbeam of polarized red light vibrating *parallel* to the cell' long axis. On the basis of the previous experiment, we can predict that red light vibrating parallel to the cell's long axis will induce chloroplast rotation.

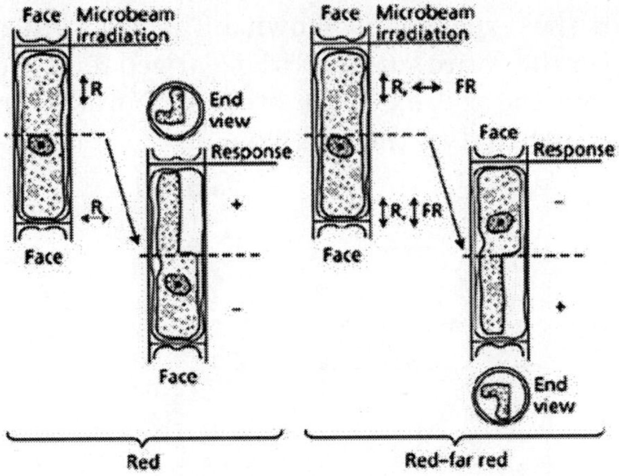

However, the red light treatment was immediately followed by a beam of far-red light. One half of the cell (top) was given a second irradiation of far-red light vibrating *perpendicular* to the cell axis; the other half of the cell (bottom) was given a second irradiation of far-red light vibrating *parallel* to the cell axis. As shown in the right-hand side of Figure above, far-red light prevented red light–induced chloroplast rotation only when it was vibrating perpendicular to the cell axis. This result was explained by the proposal that phytochrome undergoes a 90° change in orientation upon photoconversion of Pr to Pfr and vice versa.

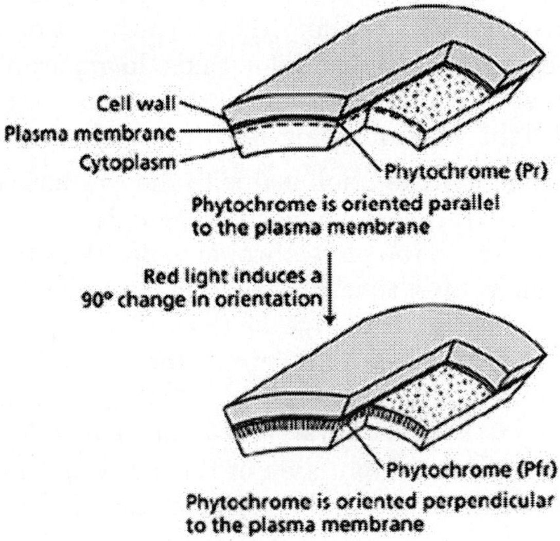

A Model for Chloroplast Rotation

Finally, what is the mechanism of chloroplast rotation in *Mougeotia*? The effects already described are best accounted for by a model in which phytochrome is located either on the plasma membrane of *Mougeotia* or on the microtubules that form a cylindrical scaffold directly beneath the membrane. It is difficult to imagine how Pr and Pfr could maintain their respective orientation without being associated either with the membrane or with the cylinder of microtubules.

The nucleotide sequence that codes for *Mougeotia* phytochrome, consistent with those of other known phytochromes, suggests that the phytochrome is not an integral membrane protein, because it lacks hydrophobic transmembrane domains. On the other hand, the amino acid sequence at the carboxy-terminal end is similar to that of proteins that bind microtubules. So the evidence suggests that phytochrome is bound to the cytoskeleton adjacent to the plasma membrane.

During rotation, the chloroplast moves away from a localized region of Pfr and toward a region of Pr. *Mougeotia* chloroplasts move in response to unilateral light as a result of gradients of the active form of phytochrome, Pfr, that form an alternating pigment pattern around the periphery of the cell cylinder, as part B of the figure shows. The gradients of Pfr are rapidly translated into gradients of actin–myosin microfilament interactions, which generate the motive force for chloroplast rotation.

The primary mechanism of Pfr action remains obscure. Calcium ions from internal stores and calcium-activated calmodulin are known to participate in the response by interacting with the scaffold of microtubules, but the precise relationship between phytochrome and calcium is not yet understood. Studies using drugs that inhibit microtube and microfilament action do not support a role for microtubules in the mechanism of chloroplast movement. However, experiments with Cytochalasin B do indicate that microfilaments are involved.

Regulation of Transcription by *Cis*-Acting Sequences

The precise number and type of *cis*-acting sequences involved in regulating transcription of a given gene have been determined by various approaches:

Conserved sequences: The sequences of the promoter regions of a given gene (for example, *RBCS*) are first compared within members of the gene family in a particular species, and between gene family members of different species. Sequences acting as *cis*-acting regulatory elements are often represented in many species (that is, they are conserved) because they are essential for function, while the DNA sequences around them may be less essential, and therefore more variable. This approach gives an idea of the locations of sequences that may be involved in regulating the expression of a gene. For *RBCS* genes, the regulatory element typically encompasses a region about 1 kilobase of DNA upstream from the transcriptional start site.

Identification of response elements: Once general locations of the conserved sequences have been identified, reporter constructs of the type already described can be tested to determine whether this fragment is necessary and sufficient to confer light regulation on the expression of the reporter sequence. Then increasing amounts of the 52 end of the promoter region can be deleted to determine the minimum piece of DNA required for expression. Typically, 250 base pairs of DNA upstream of the transcription start site are required for light-regulated transcription. Internal deletions or the insertion of small, unrelated sequences, called *linkers*, can be used to define *cis*-acting regulatory elements. Conferring light regulation. Synthetic copies of putative elements that contain one or more copies of a sequence can be inserted into non-light-regulated promoters to see if they can confer light regulation.

Identification of trans-acting factors: In vitro experiments in which nuclear protein extracts are allowed to bind to these different DNA fragments can be used to characterize the protein factors that bind to the *cis*-acting elements. Together, these four approaches have yielded detailed promoter maps for several genes, including *RBCS* and oat *PHYA*. Some of the important *cis*-acting elements present in many light-regulated genes include GT-1 sites (GGTTAA), I boxes (GATAA) and G boxes (CACGTC).

The *trans*-acting factors that bind to these three elements have also been characterized. However, each of these three *cis*-acting regulatory elements has also been found in the promoters of some non-light-regulated genes. Moreover, none of these elements alone is able to confer light regulation onto a non-light-regulated gene. Thus, light regulation arises from various combinations of different general elements rather than from a single regulatory element. Recent evidence suggests that the minimal sequence required for light-inducible gene expression is a pairwise combination of two elements such as GT-1 sites, G boxes, or I boxes.

Phytochrome Regulation of Gene Expression

We have seen that phytochrome-mediated light responses at the cellular level can involve rapid effects on membrane properties. The stimulation and repression of transcription by light can also be very rapid, with lag times as short as 5 minutes. Such early-gene expression is likely to be regulated by the direct interaction of transcription factors by one or more phytochrome-initiated signal transduction pathways. Within the nucleus, these transcription factors stimulate the transcription of specific genes. Some of the gene products of the early genes are transcription factors themselves, which activate the expression of other genes. Expression of the early genes, also called primary response genes, is independent of protein synthesis; expression of the late genes, or secondary response genes, requires the synthesis of new proteins.

As with cytokinin-regulated gene expression, the photoregulation of gene expression has focused on the nuclear genes that encode messages for chloroplast proteins: the small subunit (SSU) of ribulose-1,6-bisphosphate carboxylase/oxygenase (rubisco) and the major light-harvesting chlorophyll *a*/*b*–binding proteins associated with the light-harvesting complex of photosystem II (LHCIIb proteins). These proteins play important roles in chloroplast development and greening; hence their regulation by phytochrome has been studied in detail. The genes for both of these proteins—*RBCS* and *LHCB* (also called *CAB* in some studies)—are present in multiple copies in the genome.

The abundance of these mRNAs has been shown to increase in response to light and phytochrome action. The abundance of other mRNAs, however, can be decreased by phytochrome action; this latter class includes mRNAs that encode oat phyA and the chlorophyll biosynthesis enzyme NADPH–protochlorophyllide oxidoreductase. Although the abundance of most of these mRNAs is regulated by light given in the low-fluence range, some *LHCB* mRNAs also respond to light in the very-low-fluence range.

We can demonstrate phytochrome-regulated induction of mRNA abundance (for example *RBCS* mRNAs) experimentally by giving etiolated plants a brief pulse of low-fluence red or far-red light, returning them to darkness to allow the signal transduction pathway to operate, and then measuring the abundance of specific mRNAs in total RNA prepared from each set of plants. If its abundance is regulated by phytochrome, the mRNA is absent or present at low levels in etiolated plants, but is increased by red light. The red light–

induced increase in expression can be reversed by immediate treatment with far-red light, but far-red light alone has little effect on mRNA abundance.

In the case of mRNAs (e.g., *LHCB*) that respond to very-low-fluence red light, expression increases in response to both red and far-red light, and additional experiments are needed to confirm the involvement of phytochrome. In the case of mRNAs whose expression is negatively regulated by phytochrome, for example oat *PHYA* mRNA, levels are *high* in darkness, *reduced* by a pulse of red light, and *restored* to dark levels by red/far-red light treatment.

However, none of these experiments prove the action of phytochrome on transcription; instead they show the effect of phytochrome only on overall mRNA abundance in total RNA, which reflects both transcription and degradation processes. As mentioned earlier, the abundance of oat *PHYA* mRNA in total RNA is the result of both transcription and degradation events.

The involvement of phytochrome in transcriptional regulation of mRNA abundance has been demonstrated by two approaches. In the first approach, etiolated plants are treated as was described for the mRNA studies, and then transcriptionally active nuclei are isolated and allowed to elongate transcripts initiated under the different in vivo light treatments in the presence of radioactive precursors.

If the relative incorporation into an mRNA is similar to that described for the overall abundance in total RNA, then transcription of the mRNA is thought to be under phytochrome control. The second approach involves fusing the promoter of the gene to be tested with the coding sequence of a reporter gene such as *GUS*. By placing the fused gene back into the plant it is possible to measure the abundance of the reporter mRNA or its encoded enzyme activity. Both of these approaches have been used extensively, and the latter has allowed the definition of the minimum amount of the upstream promoter region required to confer phytochrome-regulated transcription of the linked reporter sequence.

Recently, Elaine Tobin and her colleagues at the University of California, Los Angeles, characterized a MYB-related transcription factor whose mRNA level increases rapidly when *Arabidopsis* is transferred from the dark to the light. This DNA-binding protein appears to bind to the promoter of an *LHCB* gene and regulate its transcription,, occurs later than the increase in the MYB-related protein. The gene that encodes the MYB-related protein is therefore probably a primary response gene, while the *LHCB* gene is probably a secondary response gene.

More recently, microarray analysis has been applied to the study of light signal transduction. In microarray analysis short oligonucleotides or cDNA sequences representing thousands of gene sequences are spotted on a glass slide. RNA samples isolated from control and experimental plants are labelled using two different fluorescent dyes. Both the control and the experimental sample are hybridized to the microarray and the slide is then scanned using two lasers that will excite one or the other dye. The light emitted from the dye is proportional to the amount of RNA that is bound to the gene-specific oligonucleotide or cDNA sequence and thus provides a relative measure of gene expression. To visualize the relative expression levels from both samples, images are often false coloured so that one signal is represented with green and the other signal as a red, when the signals overlap they are yellow. The commercially available Affymetrix arrays utilizes small 25-mer oligos as probes on the glass slide that are hybridized to a single labelled sample. This technique has been used to examine changes in gene expression when plants are grown under red or far-red light to define phytochrome signalling pathways and have shown that multiple transcription factors are early targets of phytochrome control.

Two-Hybrid Screens and Co-Immunoprecipitation

Two-hybrid library screens are based on the finding that many transcriptional activators contain separable DNA binding and transcriptional activation domains. In a two-hybrid screen, a DNA binding domain of a transcription factor is fused to a test or "bait" protein such as phyA. Yeast cells carrying the bait protein are transformed with a library of plant cDNA products that are fused to a transcriptional activation domain ("prey"). If the bait protein interacts with the prey protein, then the activation domain will be brought into close proximity to the DNA binding site of the transcriptional activator, thus driving reporter gene expression that will pigment the yeast cells or provide an essential amino acid, permitting growth.

Co-immunoprecipitation experiments are often conducted to provide independent verification of the results of two-hybrid screens. These assays can be performed using *in vitro* synthesized versions of bait and prey proteins. The proteins are mixed together and immunoprecipitated using an antibody that is specific to either the bait or prey construct. If the two proteins co-precipitate, it is evidence they interact with each other.

The Roles of G-Proteins and Calcium in Phytochrome Responses

Early experiments with intact seedlings and cultured cells indicated that the control of gene expression by phytochrome involves the action of a heterotrimeric G-protein and the calcium-binding regulatory protein calmodulin.

Subsequently, a single-cell system has been developed that can be microinjected with various proposed components of the pathway and monitored for changes in gene expression. Although these experiments have not yet led to the identification of specific pathway components, they have been informative about the involvement of G-proteins and calmodulin.

Microinjection of phytochrome (mainly phyA) purified from etiolated oat seedlings into single cells prepared from the tomato *aurea* mutant (a chromophore biosynthetic mutant that has reduced levels of all phytochromes) results in stimulation of chloroplast development, accumulation of anthocyanin pigment, and expression of a GUS-coding sequence fused to a phytochrome-regulated promoter sequence. Etiolated seedlings of the *aurea* mutant show poor induction of these responses when transferred to light, probably because of the reduction in the levels of functional phyA. Inhibitors of G-proteins block the phytochrome-stimulated events in the single-cell system, while activators or intermediates stimulate the response. Increased levels of Ca^{2+} in the medium also stimulate *GUS* expression and chloroplast development when calmodulin is present, but they do not affect anthocyanin biosynthesis.

In the *Aurea* cell system, cGMP has been shown to act as a signalling intermediate in at least two signal transduction pathways. Specifically, the pathway that activates anthocyanin biosynthesis requires cGMP. Inhibitors of cGMP action inhibit this pathway; addition of cGMP activates it. In contrast, the pathway that requires calcium to activate chloroplast development and gene expression has a variable requirement for cGMP. Activation of genes that encode several chloroplast proteins, including RBCS and LHCB, does not require cGMP, while activation of genes that encode some photosystem I polypeptides and the cytochrome *b6/f* complex does.

The difference in signalling requirements is due in part to the different *cis*-acting regulatory elements found upstream of these genes. Specific DNA elements upstream of *RBCS* genes appear to be the targets for the calcium pathway, while other distinct elements located upstream of anthocyanin biosynthesis genes are the targets for cGMP. The differential regulation of the expression of phytochrome-regulated genes can now be explained in terms of their *cis*-acting regulatory elements, different combinations of which afford both tissue-specific and light-specific expression.

Phytochrome Effects on Ion Fluxes

Phytochrome can rapidly alter the properties of membranes. We have already seen that low-fluence red light is required before the dark period to induce rapid leaflet closure during nyctinasty, and that fluxes of K^+ and Cl^- into and out of extensor and flexor cells mediate the response. However, the rapidity of leaf closure in the dark (lag time about 5 minutes) would seem to rule out mechanisms based on gene expression. Instead, rapid phytochrome-induced changes in membrane permeability and transport appear to be involved.

The role of the plasma membrane proton pump in *Samanea* pulvini (the structures at the base of the petiole that include the extensor and flexor cells) was studied by use of a proton-sensitive liquid membrane electrode. During phytochrome-mediated leaflet closure, the apoplastic pH of the flexor cells (the cells that swell during leaflet closure) decreased, while the apoplastic pH of the extensor cells (the cells that shrink during leaflet closure) increased. Thus, the plasma membrane H^+ pump of the flexor cells appears to be activated by darkness (provided that phytochrome is in the Pfr form), whereas the H^+ pump of the extensor cells appears to be deactivated under the same conditions.

Increased H^+ pump activity on the plasma membranes of the flexor cells should increase the membrane potential, thereby allowing more K^+ to enter the cells. Conversely, a reduction in the plasma membrane H^+ pump activity in the extensor cells would promote the loss of K^+ ions. Consistent with this model, the reverse pattern of apoplastic pH change was seen during leaflet opening.

Studies have also been carried out on phytochrome regulation of K^+ channels. Since it is difficult to study membrane channels in the intact pulvinus, Richard Crain and his colleagues at the University of Connecticut isolated protoplasts (cells without their cell walls) of both extensor and flexor cells from *Samanea* leaves. They determined the open or closed state of the K+ channels indirectly by measuring the membrane potential changes that occurred when the extracellular K^+ concentration was raised from 20 to 200 m*M*. When the extracellular K^+ concentration was raised, K^+ entered the protoplasts and depolarized the membrane potential only if the K^+ channels were open. When the extensor and flexor protoplasts were transferred to constant darkness, the state of the K^+ channels exhibited a circadian rhythmicity during a 21-hour incubation period and the two cell types varied reciprocally, just as they do in vivo. That is, when the extensor K^+ channels were open, the flexor K^+ channels were closed, and vice versa. Thus, the circadian rhythm of leaf movements has its origins in the circadian rhythm of K^+ channel opening.

To test the effects of phytochrome on K^+ channels, researchers first treated protoplasts with white light, then exposed them to a brief pulse of either red light alone or red light followed by far-red light and transferred them to continuous darkness. Red light caused the K^+ channels of the flexor protoplasts to open, and far-red light reversed the effect of red light. The extensor protoplasts were closed by both treatments. In contrast to red light, blue light caused the K^+ channels of the extensors to open, and those of the flexors to close. Thus, phytochrome seems to regulate specifically the K^+ channels of the flexor cells. Blue light regulates primarily the extensor cells, although it can also affect the flexors at specific times of day.

On the basis of the evidence thus far, we can conclude that phytochrome brings about leaflet closure by regulating the activities of the primary proton pumps and/or the K+ channels of the flexor and extensor cells. Although the effect is rapid, it is not instantaneous, and it is therefore unlikely to be due to a direct effect of phytochrome on the membrane. Instead, phytochrome acts indirectly via one or more signal transduction pathways, as in the case of the regulation of gene expression by phytochrome. However, some effects of red and far-red light on the membrane potential are so rapid that phytochrome may also interact directly with the membrane. Such rapid modulation has been measured in individual cells and has been inferred from the effects of red and far-red light on the surface potential of roots and oat (*Avena*) coleoptiles. The lag between the production of Pfr and the onset of measurable potential changes varies from organism to organism; for example, it is about 1.7 s for depolarization (inside becomes less negative relative to outside the plasma membrane) in the giant alga *Nitella* and 4.5 s for hyperpolarization in *Avena*.

Changes in the bioelectric potential of cells imply changes in the flux of ions across the plasma membrane. The *Mougeotia* microbeam irradiation experiments of Haupt seemed to support the idea that phytochrome is localized on the plasma membrane, and membrane isolation studies carried out in the 1970s provided evidence that a small portion of the total phytochrome is tightly bound to various organellar membranes. These findings led some workers to suggest that membrane-bound phytochrome represents the physiologically active fraction, and that all the effects of phytochrome on gene expression are initiated by changes in membrane permeability. On the basis of sequence analysis, however, it is now clear that phytochrome is a hydrophilic protein without membrane-spanning domains. The current view is that phytochrome in *Mougeotia* is associated with microtubules located directly beneath the plasma membrane.

If phytochrome exerts its effects on membranes from some distance, no matter how small, involvement of a "second messenger" would be implied. For example, rapid changes in cytosolic free calcium have been implicated as second messengers in several signal transduction pathways, and there is evidence that calcium plays a role in chloroplast rotation in *Mougeotia*. More recently, however, the role of calcium as a second messenger in phytochrome action has been questioned, so calcium-independent mechanisms may also be involved.

Microarray Analysis of Shade Avoidance

To identify components of the shade-avoidance response pathway, microarray analysis was conducted in Arabidopsis, comparing plants grown in low and high R:FR light environments. These studies have identified both early-and late-acting genes that regulate responses to shading by other plants. Characterizations of two Arabidopsis transcription factors (*ATHB-2* and *PIL1*) also suggests that shade avoidance is initiated by rapid transcriptional changes following a decrease in the ratio of red to far-red light. These profiling studies strengthen the previous genetic studies that show that phyA acts to antagonize phyB-mediated responses under prolonged low R:FR light conditions through the transcriptional control of target genes that include components of the auxin transport and signalling pathways. Interestingly, it appears that the circadian clock is capable of regulating or "gating" physiological responses to R:FR through a modulation of gene expression. That is, changes in gene expression due to a change in the R:FR only occur at certain phases of the circadian rhythm.

Blue-Light Responses: Stomatal Movements and Morphogenesis

Guard Cell Osmoregulation and a Blue Light-Activated Metabolic Switch

Studies on photosynthetic algae, such as *Chlorella,* have shown that irradiation by blue light increases respiratory rates of dark-grown cells, with an action spectrum that is typical of specific blue-light responses. These changes in respiratory rates are associated with a drastic shift in metabolism, which can also be observed in experiments using blue and red light.

Red light stimulates photosynthetic carbon fixation in these algae, and the fixed carbon is incorporated into glucose and sucrose. The metabolic shift is observed when low photon fluxes of blue light are added to the red-light treatment; glucose and sucrose are no longer synthesized, and the fixed carbon is used in the biosynthesis of organic acids, amino acids, and proteins.

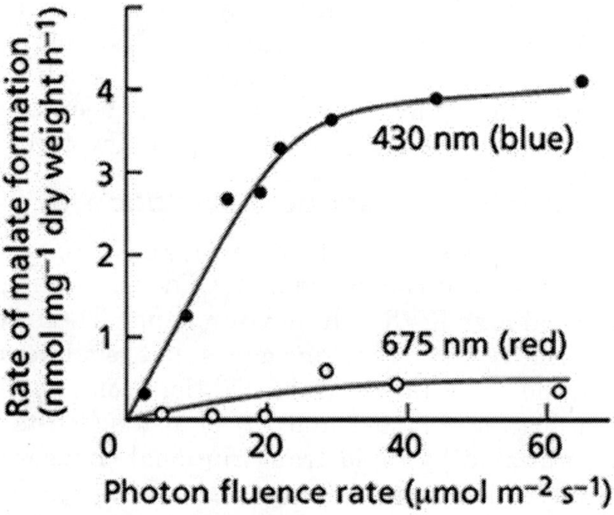

Figure: *Blue light stimulates malate synthesis in guard cells. Under red light, guard cells synthesize mostly glucose and sucrose; this metabolic pathway is deactivated by blue light, and metabolism switches to starch degradation and malate biosynthesis.*

Blue light is also a powerful regulator of guard cell metabolism. Guard cells irradiated with red light accumulate sucrose; if low photon fluxes of blue light are added to the red-light irradiation, sucrose stops accumulating, and guard cells accumulate K^+ and malate^{2-}. Thus, the blue light–stimulated shift to the synthesis of the organic acid malate in guard cells parallels the shift to the synthesis of organic acids and amino acids in photosynthetic algae.

The experimental use of blue and red light has been useful for investigations of osmoregulatory pathways associated with stomatal movements. The rapid activation of proton pumping by blue light generates a proton gradient at the guard cell plasma membrane that drives potassium and chloride uptake. Blue light also stimulates starch hydrolysis and malate biosynthesis. Prolonged incubations of guard cells in detached epidermis under low fluence rates of blue light result in large stomatal apertures accompanied by extensive starch hydrolysis.

Guard cells also show a sucrose-dependent osmoregulatory phase. Sucrose can originate from starch hydrolysis. On the other hand, guard cells from detached epidermis irradiated with red light under ambient CO_2 concentrations operate a third metabolic pathway, showing no detectable potassium uptake or starch hydrolysis. In these conditions, guard cells slowly accumulate sucrose, and this sucrose accumulation is blocked by DCMU (dichlorophenyldimethylurea), an inhibitor of photosynthetic electron transport. A recent study of guard cell osmoregulation under red light illumination and CO_2-free air showed that under these conditions, in which photosynthesis is inhibited by the absence of CO_2, red light activates the potassium-dependent pathway.

These studies underscore the remarkable metabolic flexibility of guard cells. A study of the effects of calcium and abscisic acid on stomatal apertures under red and blue light found that increasing calcium or ABA concentrations inhibited blue light–stimulated stomatal opening in a concentration-dependent fashion, but had no effect on red light–stimulated opening.

These contrasting responses to blue and red light can be explained by the effect of the treatments on guard cell osmoregulation. ABA and high calcium concentrations have been shown to inhibit proton pumping and potassium uptake, which are central to blue light–stimulated stomatal opening. Red light, on the other hand, stimulates guard cell photosynthesis and sucrose accumulation, and this osmoregulatory pathway appears to be insensitive to ABA and calcium. Unlike visual systems, the colour information in the blue light and red light treatments does not appear to have functional significance for photosynthetic organisms. Rather, their dramatic experimental effects appear to be a result of a differential stimulation of carotenoids and chlorophylls that are selectively excited by blue light applied under a saturating background of red light, or by red light alone, respectively. Under solar radiation, regulatory pathways activated by the excitation of these pigments are used by the chloroplast to sense key features in their environment and to control their metabolism.

The Coleoptile Chloroplast

Guard cell chloroplasts are a conspicuous structural feature of guard cells, and their physiology and biochemistry have been studied for several decades. In contrast, the chloroplasts of phototropically sensitive coleoptiles have only recently been analysed in detail.

The coleoptile chloroplast had been suggested as a possible site for photoreception in oat coleoptiles by Kenneth V. Thimann and his coworkers in 1960, but interest in the flavin hypothesis later in that decade shifted attention away from coleoptile chloroplasts and these organelles have been virtually ignored for several decades.

Recent studies have shown that the coleoptile chloroplast evolves oxygen, fixes CO_2 photosynthetically, and operates the xanthophyll cycle. Dark-grown, etiolated coleoptile tips have violaxanthin and antheraxanthin but no zeaxanthin, and they accumulate zeaxanthin when irradiated.

When dark-grown coleoptiles are pretreated with red light for different lengths of time, their zeaxanthin content is proportional to the length of the red-light treatment. Exposure of coleoptiles with different zeaxanthin content to a brief pulse of blue light elicitis phototropic bending that is linearly related to the zeaxanthin content of the coleoptiles. Most interestingly, the regression line through the data points extrapolates to zero, suggesting that there is no phototropic response in the absence of zeaxanthin.

Red light–treated coleoptiles with a high zeaxanthin content convert their zeaxanthin to violaxanthin when transferred to the dark. Phototropic bending in response to a pulse of blue light given at different time intervals after the transfer to the dark is linearly related to zeaxanthin content.

When corn coleoptiles are given a saturating red-light pretreatment and a blue-light pulse in the presence of increasing concentrations of DTT (an inhibitor of zeaxanthin formation), bending and zeaxanthin content decrease linearly as a function of DTT concentration.

Thus the relationship between the zeaxanthin content of coleoptile chloroplasts and blue light–dependent phototropic sensitivity is very similar to the relationship between the zeaxanthin content of the guard cell chloroplast and the blue-light sensitivity of stomatal opening.

Coleoptile and guard cell chloroplasts share some properties that are clearly distinct from mesophyll chloroplasts, including high rates of oxygen evolution and low rates of photosynthetic carbon fixation, a high sensitivity of their xanthophyll cycle to low levels of incident photon fluxes, and a high starch content. The coleoptile and guard cell chloroplasts also show a specific blue light response with an action spectrum that closely matches the action spectra for blue light–stimulated stomatal opening and coleoptile phototropism.

These remarkable similarities suggest that these distinct characteristics are typical of chloroplasts specialized in signal transduction, as opposed to the classic role of the mesophyll chloroplast, specialized in photosynthetic carbon fixation.

Chapter

7

Abscisic Acid: A Seed Maturation and Antistress Signal

Structural Requirements for Biological Activity of Abscisic Acid

The features shown to be essential for biological activity include the carboxyl group, the tertiary hydroxyl group, the 2-*cis*, 4-*trans*-pentadienoic side chain, a 4'-ketone, and a double bond in the cyclohexane ring. Catabolic products of ABA that are present in the tissue and are missing any of these groups are biologically inactive.

The Bioassay of ABA

Coleoptile growth, the classic bioassay devised for auxins, is also used for ABA detection in plant extracts by measurement of coleoptile growth inhibition. This bioassay has adequate sensitivity (minimum detectable level is 10^{-7} *M*) and shows a linear response in the range of 10^{-7} to 10^{-5} *M*, but it has some disadvantages. Plant extracts have other promoters and inhibitors, which decrease the specificity of the bioassay and make preliminary purification steps necessary. Other available bioassays for ABA include inhibition of germination, inhibition of gibberellic acid– induced α-amylase synthesis in cereal aleurone layers, induction of leaf abscission, and stomatal closure. Stomatal closure is highly specific for ABA because it is affected little by other plant growth regulators. Additional advantages of this bioassay include a fast response of guard cells to ABA, high sensitivity (minimum detectable level 10^{-9} *M*), and a linear response over a wide range of concentrations.

Proteins Required for Desiccation Tolerance

LEA protein sequences are highly conserved among a wide variety of species. They are related to two other groups of proteins, the RAB (*r*esponsive to *AB*A) and DHN (*de*h*y*dri*n*) proteins. These protein classes accumulate in a variety of tissues experiencing cellular dehydration, e.g. maturing embryos and pollen, and vegetative tissues exposed to drought stress, including extreme desiccation of "resurrection plants" (*Craterostigma plantagineum*).

The LEA, RAB, and DHN proteins are all water soluble, basic, rich in glycine and lysine, and low in hydrophobic residues. As a result they are extremely hydrophilic and stable to boiling. These features led Dure and colleagues (1989) to propose that they function specifically in the protection of membranes and proteins against desiccation damage, possibly by binding water tightly or providing hydrophilic interactions in the absence of free water, and by preventing crystallization of cellular components through their ability to act as stabilizing "solvents," similar to the effects of sugars. One of these proteins confers increased osmotolerance in transgenic yeast.

Types of Seed Dormancy and the Roles of Environmental Factors

During seed maturation, the embryo enters a quiescent phase in response to desiccation. Seed germination can be defined as the resumption of growth of the embryo of the mature seed; it depends on the same environmental conditions as vegetative growth does. Water and oxygen must be available, the temperature must be suitable, and there must be no inhibitory substances present.

However, in many cases a viable (living) seed will not germinate even though all the necessary environmental conditions for growth are satisfied. This phenomenon is termed seed dormancy. Seed dormancy introduces a temporal delay in the germination process that provides additional time for seed dispersal over greater geographical distances. It also maximizes seedling survival by preventing germination under unfavourable conditions.

Two types of seed dormancy have been recognized, coat-imposed dormancy and embryo dormancy. Coat-imposed dormancy is dormancy imposed on the embryo by the seed coat and other enclosing tissues, such as endosperm, pericarp, or extrafloral organs. The embryos of such seeds will germinate readily in the presence of water and oxygen once the seed coat and other surrounding tissues are either removed or damaged. There are five basic mechanisms of coat-imposed dormancy:

Prevention of water uptake. Prevention of water uptake by the seed coat is a common cause of seed dormancy in families found in arid and semiarid regions, especially among legumes, such as clover (*Trifolium* spp.) and alfalfa (*Medicago* spp.). Waxy cuticles, suberized layers, and lignified sclereids all combine to restrict the penetration of water into the seed.

Mechanical constraint. The first visible sign of germination is typically the radicle breaking through the seed coat. In some cases, however, the seed coat may be too rigid for the radicle to penetrate. Nuts with hard, lignified shells are examples of dormancy caused by mechanical constraint. Such shells must be broken by biotic or environmental forces for the seed to germinate. Even nonlignified tissues, such as the endosperm of lettuce seeds, can suppress expansion of the embryo. For the seeds to germinate, the endosperm cell walls must be weakened by the production of cell wall–degrading enzymes.

Interference with gas exchange. Some seed coats are considerably less permeable to oxygen than an equivalent thickness of water is—e.g., less by a factor of about 10^4 in seeds of cocklebur (*Xanthium pennsylvanicum*). This lowered permeability to oxygen suggests that the seed coat inhibits germination by limiting the oxygen supply to the embryo. In support of this idea, investigators can break the dormancy of such seeds either by making a small hole in the coat with a pin (without weakening the coat mechanically), or by treating the coat with concentrated oxygen.

However, other studies suggest that the oxygen consumption of the embryos from such seeds is considerably less than the amount of oxygen able to penetrate the seed coats under normal aerobic conditions. Thus the role of oxygen impermeability in seed coat dormancy remains unresolved.

Retention of inhibitors. The seed coat may prevent the escape of inhibitors from the seed. For example, growth inhibitors readily diffuse out of isolated *Xanthium* embryos, but not from intact seeds.

Inhibitor production. Seed coats and pericarps may contain relatively high concentrations of growth inhibitors that can suppress germination of the embryo. ABA is a common germination inhibitor present in these maternal tissues. In certain cases where repeated washing (leaching) removes dormancy, the effect is thought to be due to the loss of such inhibitory compounds.

The second type of seed dormancy is embryo dormancy, which refers to a dormancy that is inherent in the embryo and is not due to any influence of the seed coat or other surrounding tissues. In some cases, embryo dormancy can be relieved by amputation of the cotyledons. Species in which the cotyledons exert an inhibitory effect include European hazel (*Corylus avellana*) and European ash (*Fraxinus excelsior*). A fascinating demonstration of the cotyledon's ability to inhibit growth is found in species (e.g., peach) in which the isolated dormant embryos germinate but grow extremely slowly to form a dwarf plant. If the cotyledons are removed at an early stage of development, however, the plant abruptly shifts to normal growth.

Embryo dormancy is thought to be due to the presence of inhibitors, especially ABA, as well as the absence of growth promoters, such as GA (gibberellic acid). The loss of embryo dormancy is often associated with a sharp drop in the ratio of ABA to GA.

Seed Dormancy May be Primary or Secondary

Different types of seed dormancy also can be distinguished on the basis of the timing of dormancy onset rather than the cause of the dormancy.

- Seeds that are released from the plant in a dormant state are said to exhibit primary dormancy.
- Seeds that are released from the plant in a nondormant state but which become dormant if the conditions for germination are unfavourable exhibit secondary dormancy. For example, seeds of *Avena sativa* (oat) can become dormant in the presence of temperatures higher than the maximum for germination, whereas seeds of *Phacelia dubia* (small-flower scorpionweed) become dormant at temperatures below the minimum for germination. The mechanisms of secondary dormancy are poorly understood.

Environmental Factors Control the Release from Seed Dormancy

Various external factors release the seed from dormancy, and dormant seeds typically respond to more than one condition. Many seeds lose their dormancy when their moisture content is reduced to a certain level by drying. This method of breaking seed dormancy is called afterripening, and is usually performed in a special drying oven. On the other hand, if the seed becomes too dry (5% water content or less), the effectiveness of afterripening is diminished.

Another factor that can release seeds from dormancy is low temperature, or chilling. Many seeds require a period of cold (0 to 10°C) while in a fully hydrated (imbibed) state in order to germinate. In temperate-zone species, this requirement is of obvious survival value, since such seeds will germinate not in the fall, but only in the following spring. Chilling seeds to break their dormancy is a time-honored practice in horticulture and forestry and traditionally has been referred to as *stratification*. This term is derived from the old agricultural practice of allowing seeds with a chilling requirement to overwinter outdoors in layered mounds of moist soil. Today the seeds are simply stored in a refrigerator.

The third external factor that plays an important role in breaking seed dormancy is light. Many seeds have a light requirement for germination, which may involve only a brief exposure, as in the case of lettuce, an intermittent treatment (e.g., succulents of the genus *Kalanchoe*), or even a specific photoperiod involving short or long days. The main sensor for light-regulated seed germination. Interestingly, all light-requiring seeds exhibit seed coat dormancy, and removal of the outer tissues of the seed allows the embryo to germinate in the absence of light. The effect that light has on the embryo is thus to enable the radicle to penetrate the seed coat. This penetration often involves some enzymatic weakening of the enclosing tissues.

The Longevity of Seeds

How long can a seed remain dormant and still remain viable? Seed longevity is of practical importance because of ongoing efforts to preserve plant genetic resources for future agricultural crops by setting up seed gene banks. Spectacular claims of dormant seeds that remained viable for thousands of years have been made, but are considered highly controversial. Extreme examples of ancient seeds that have been reported to retain their viability include submerged lotus seeds found in a 3,000-year-old boat near Tokyo, barley seeds from the 3,000-year-old tomb of King Tutankhamen, and arctic lupine seeds associated with rodent burrows determined to be 14,000 years old. In most cases, however, a scientifically rigorous examination of the data has either disproved or raised serious doubts about the claimed antiquity of the seeds (Bewley and Black 1994).

The most serious criticism is that the dates of the seeds have been inferred from the age of their immediate surroundings or associated artifacts rather than from direct measurements of the seeds themselves. This question raises the possibility that the seeds are much younger than the site at which they were found.

In fact, microscopic examination of authenticated ancient cereal grains from various Middle Eastern sites has shown that cells and tissues of these seeds are no longer intact. Such seeds are clearly inviable.

Although the most sensational claims for seed longevity are almost certainly bogus, seeds of *Canna compacta* apparently can live for at least 600 years. Viable *Canna* seeds were obtained from inside a walnut in a tomb in Argentina. The *Canna* seeds had apparently been inserted into the immature seeds of a growing walnut fruit before the hard outer shell formed. Once the shell hardened and the nut dried out, the result was a rattle. Native people strung the rattles together to form a necklace. In this case, the seeds had to be at least as old as the walnut shell, and carbon dating of the shell indicated that it was about 600 years old.

Plant collections begun during the late eighteenth century in Europe—for example, at the British Museum in London and at the Museum of Natural History in Paris—have been reliable sources of seeds for viability determinations. *Cassia multijuga* (false sicklepod) seeds, collected in 1776, were still viable after being tested in 1934.

In 1879, W. J. Beal initiated the longest-running experiment on seed longevity by burying the seeds of 21 different species in unstoppered bottles in a sandy hilltop near the Michigan Agricultural College in East Lansing. After 100 years, only one species, *Verbascum blattaria* (moth mullein), remained viable. Why are some seeds able to survive in a dormant state for hundreds of years, whereas others lose viability in less than 5 years? Almost nothing is known about the mechanisms that determine the longevity of seeds.

If these mechanisms could be understood, we might someday be able to greatly increase the seed longevity of agriculturally important species and varieties, thereby enhancing our ability to preserve plant genetic resources for generations to come.

Genetic Mapping Of Dormancy: Quantitative Trait Locus (QTL) Scoring of Vegetative Dormancy Combined with a Candidate Gene Approach

Complex traits, e.g. dormancy, are affected by many different, often unlinked genes. Consequently, attempts to map them result in continuous, or quantitative, variation within a mapping population.

Quantitative trait locus (QTL) mapping is used in these instances to map individual genes affecting the trait. In QTL mapping, inbred lines that differ in the trait of interest are crossed and an F_2 population is produced.

The F_2 population has continuous variation for the trait of interest, and the goal of the analysis is to determine which chromosomal segments contain genes affecting the trait. Using marker loci along each chromosome (loci that have different alleles between the parental inbred lines), the genotypes of entire chromosomal segments in each F_2 individual can be inferred as being homozygous for one or the other parental lines or heterozygous.

If the inheritance of a chromosomal segment is correlated with the trait of interest, then a gene affecting the trait is likely to be between the two marker loci. By analysing chromosomal segments along each of the chromosomes, the number and locations of genes affecting the quantitative trait can be determined.

Candidate genes are specific loci, identified by predicted biochemical function or homology to regulatory genes from other species, that are suspected of affecting the trait of interest. Molecular differences between the parental alleles at these loci can be used to determine whether they map to the same region as any of the QTLs, thereby supporting the hypothesis that a given gene contributes to regulation of the trait.

ABA-Induced Senescence and Ethylene

The ethylene has also been found to play a role in the senescence of oat leaf segments, but ABA appears to be the initiating agent, whereas ethylene appears to exert its effects at a later stage. To determine whether the ABA effects are mediated by ethylene, the effects of ABA and ethylene on senescence of leaf discs from wild-type and ethylene-insensitive mutants were compared. Chlorophyll loss from untreated leaf discs of the ethylene-insensitive mutant was much slower than that from the wild type, in the light and in the dark.

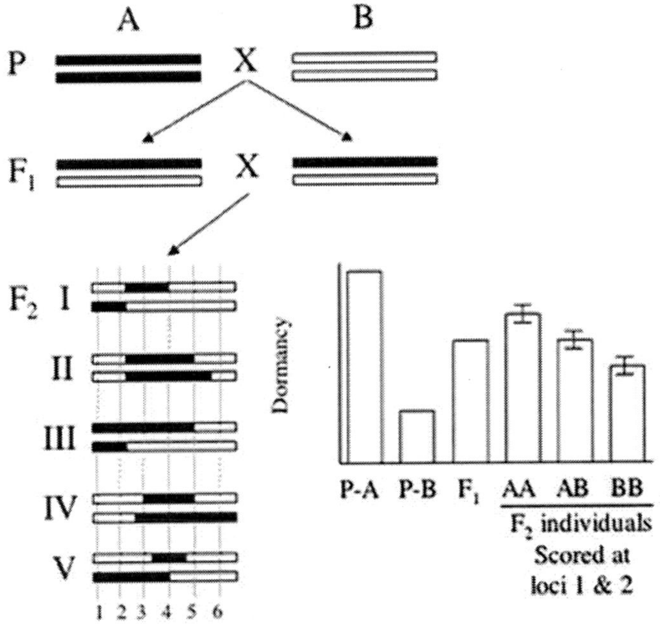

Figure: *Two parental (P) inbred lines, A & B, which differ for a trait of interest (e.g. dormancy) are crossed to produce heterozygote F_1 progeny which have an intermediate degree of dormancy. Only one chromosome pair is depicted for simplicity. The F_1 progeny are crossed to produce a large segregating F_2 population (usually 100-500 individuals are analysed, here only five individuals, I-V, are depicted). Because of recombination during F_1 meiosis, the F_2 individuals have varying genetic combinations of the original inbred lines (depicted by shading of chromosomes) and have continuous variation, from high to low, in degree of dormancy. In order to estimate whether segments of the chromosome are homozygous for one of the parental-lines or heterozygote, the F_2 individuals are scored for marker loci (1-6) along the chromosome. For example, loci 1 & 2 are identified as homozygous for A-line alleles in individual III, as heterozygous in individuals I & V and as homozygous for B-line alleles in individuals II & IV. The F_2 individuals are also scored for their dormancy. After scoring many F_2 individuals a correlation between the genotypes at locus 1 & 2 and phenotype is determined and therefore this segment likely contains at least one gene affecting dormancy (bar graph). The analysis is repeated for each chromosomal segment to identify which ones are likely to contain genes affecting dormancy.*

Ethylene accelerated the chlorophyll loss from the wild-type leaf discs, but it had no effect on the yellowing of the ethylene-insensitive mutant discs. In contrast, treatment with ABA stimulated chlorophyll loss in *both* the wild-type and the mutant discs, whether they were held in the light or in the dark (Zacarias and Reid 1990). These results indicate that the promotion of senescence by ABA is not mediated through its stimulation of ethylene production.

Yellow Cameleon: A Noninvasive Tool for Measuring Intracellular Calcium

Cytoplasmic free calcium ($[Ca^{2+}]_{cyt}$) acts as a secondary messenger in a variety of signalling pathways and it appears that the frequency, magnitude and localization of $[Ca^{2+}]_{cyt}$ oscillations are important determinants of specificity. “Traditional” techniques for measuring Ca^{2+} have made use of microinjected Ca^{2+} sensitive indicator dyes, but these methods are technically difficult and invasive. An alternative strategy makes use of Ca^{2+} binding proteins that can be transgenically expressed throughout an organism, eliminating the need for microinjection. The Ca^{2+} indicator yellow cameleon is a fusion protein composed of four domains: two modified green fluorescent proteins (GFPs) linked via calmodulin (CaM) and a CaM-binding peptide (“M13”).

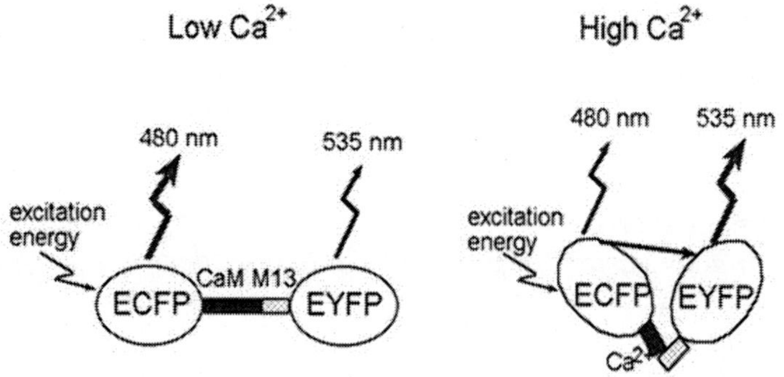

Figure: *The binding of calcium brings the two ends of the yellow cameleon protein together, altering the ratio of 535 nm to 480 nm emission.*

The GFPs are modified to shift their emission wavelengths toward blue (enhanced cyan fluorescent protein, ECFP) or yellow (enhanced yellow fluorescent protein, EYFP). When the CaM domain binds to Ca^{2+}_{cyt} it associates with the CaM-binding peptide, producing a conformational change that brings the modified GFP domains closer together. Following exposure to a laser, excitation energy from ECFP (the higher-energy fluorescent protein) is passed to EYFP resulting in increased emission by this lower-energy fluorescent protein.

Consequently, increased $[Ca^{2+}]_{cyt}$ results in increased emission at 535 nm relative to that at 480 nm, and the 535/480 emission ratio is proportional to the $[Ca^{2+}]_{cyt}$ (Miyawaki et al 1997; Allen et al. 1999).

The Two-Hybrid System

Many cellular proteins interact with a variety of other proteins to perform their physiological functions. Identification of interacting proteins can be accomplished by biochemical assays, such as co-purification of a complex, but these require relatively stable interactions and the ability to identify components of the complex immunologically or based on amino acid sequence and composition. Alternatively, the two-hybrid assay is a genetic approach that takes advantage of the modular nature of proteins to identify interacting domains.

Although there are several variations, in all cases two fusion proteins are co-expressed in a cell line that will activate expression of a reporter gene if the fusion proteins interact.

For example, the yeast transcription factor GAL4 can be separated into distinct DNA binding (BD) and transcription activation (AD) domains. If hybrid genes are constructed such that they encode fusion proteins BD-protein X and AD-protein Y, interactions between proteins X and Y will bring the GAL4 domains cose enough to reconstitute a functional transcription factor that can trans-activate a GAL4-responsive promoter.

The cell lines often contain several GAL4-responsive promoter-reporter fusions, whose activity can be scored either qualitatively or quantitatively. Qualitative scoring (+/- growth) makes it possible to screen a cDNA library for fusions encoding interacting proteins by requiring expression of a gene whose activity an complement an auxotrophy. Quantitative scoring (e.g. β-galactosidase activity from a *lacZ* reporter) permits comparisons of the relative strength of interactions among a variety of protein pairs.

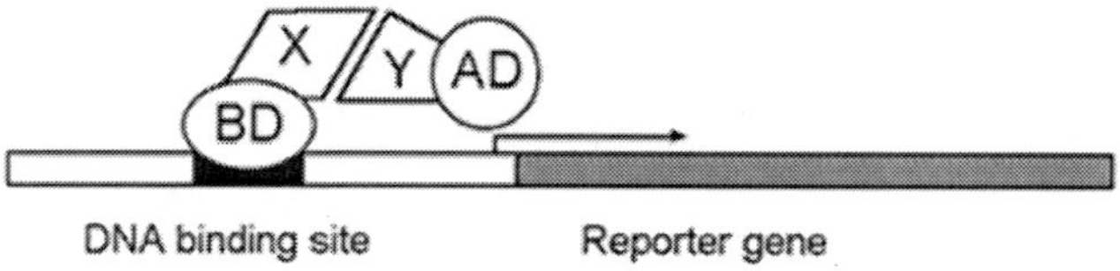

Figure: *Two fusion proteins containing the GAL4 DNA binding (BD) and transcription activation (AD) domains can reconstitute a functional transcription factor that can trans-activate a GAL4-responsive promoter if the interactions between the two proteins, X and Y, will bring the GAL4 domains close enough together.*

The Control of Flowering

Regulation of Juvenility by the TEOPOD (TP) *Genes in Maize*

The regulation of the transition from juvenile to adult (i.e., the attainment of competence) is influenced by a wide variety of conditions that affect the growth rate, and by transmissible factors such as carbohydrates and gibberellins. Hence it has been difficult to determine exactly what regulates the phase change during normal development. The analysis of mutations that affect the timing of phase changes, called heterochronic mutations, provides the opportunity to identify specific genes that regulate juvenility. The best-studied example is in maize.

The juvenile and adult phases of corn plants are characterized by differences in vegetative morphology. Scott Poethig and his colleagues at the University of Pennsylvania have been characterizing the heterochronic *teopod* (*tp*) mutants of maize: *tp1*, *tp2* and *tp3*. These semidominant mutations cause cells and tissues that would normally develop as adult structures to become more juvenile. In *tp* plants, leaves in the normally adult part of the shoot are morphologically intermediate between juvenile and adult leaves. Whereas in the wild type, tassels bear male flowers and ears bear female flowers, in *tp* mutants both tassels and ears possess leaves in addition to flowers. However, *tp* mutations affect neither the rate nor the duration of growth of the plant, nor do they interfere with the timing of tassel formation. The *teopod* mutations thus specifically affect the transition from juvenile to vegetative adult phase without affecting the transition to the reproductive phase.

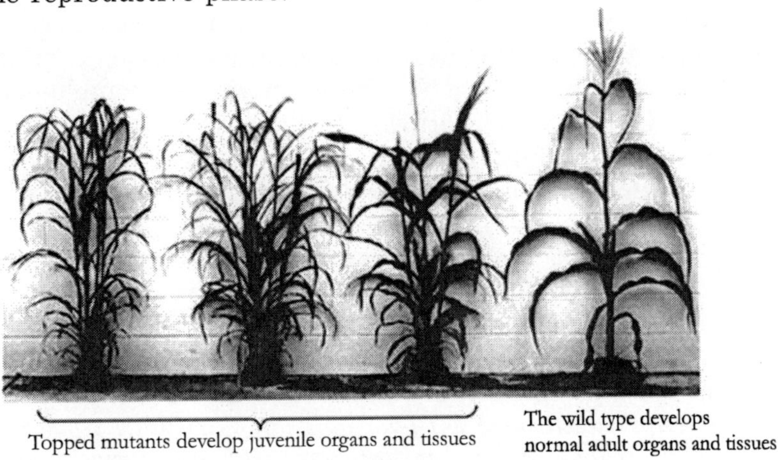

Figure: *Phenotypes of three teopod mutants and the wild-type maize plant. From left to right: tp1, tp2, tp3, and wild type.*

The possible interaction between the *TEOPOD* genes and the gibberellins has been investigated. In GA-deficient dwarf mutants of maize, the transitions from juvenile to adult vegetative development and from adult vegetative to reproductive development occur later than in wild-type plants. Thus, endogenous GAs are required for the normal timing of phase changes.

Flowering of Juvenile Meristems Grafted to Adult Plants

By definition, juvenile plants cannot be induced to flower with an inductive photoperiod, suggesting either that the shoot apical meristems of juvenile plants are not competent to flower, or that juvenile leaves produce insufficient amounts of a floral stimulus. The LSDP *Bryophyllum* passes through a juvenile phase, which lasts from 5 months to a year (depending on the species), during which it cannot be induced to flower by transfer from LD to SD.

However, when juvenile *Bryophyllum* shoots were grafted onto flowering plants, the juvenile shoots flowered promptly and profusely. This result indicates that the shoot apical meristems of juvenile plants are competent to flower, but fail to do so because the juvenile leaves produce insufficient amounts of the floral stimulus. Age is another factor that can affect production of the floral stimulus. Grafting studies with the SDP *Perilla* have shown that the leaves at the top of a plant are stronger donors of the floral stimulus than the leaves near the base of a plant, in keeping with the juvenile character of the basal leaves. The experiment was performed as follows. Two groups of *Perilla* plants were sown at different times. When the first group had produced five pairs of leaves, the second group had only produced two.

The fifth and the second pairs of leaves from the two sets of plants were grafted onto stocks and maintained under long days. Once the graft unions were established, the grafted leaves were induced with short days by daily enclosing them in opaque bags. Leaves from the fifth node needed only 14 short days to induce flowering while leaves from the second node required 28 short days. This finding shows that the more juvenile leaves from the second node produce less floral stimulus, and thus require more inductive photoperiods to induce flowering, than the fully adult leaves higher up the shoot.

Support for the Role of Blue-Light Regulation of Circadian Rhythms

The role of blue light in regulating both circadian rhythmicity and flowering is also supported by studies with an *Arabidopsis* flowering-time mutant: elf3 (early flowering *3*). The *elf3* mutant flowers earlier than the wild type under a variety of growth conditions and is insensitive to photoperiod; that is, it is day-neutral. Transformation of *elf3* with the luciferase reporter construct revealed a defect in circadian rhythmicity. The mutant displayed no circadian rhythmicity when grown in continuous light. Significantly, the *elf3* mutant also has a long hypocotyl when grown in constant white light, similar to the *hy* mutants, and is less responsive than the wild type to all wavelengths of light, but especially to blue light. This finding suggests that the circadian clock is intact in the mutant, but that the transduction of light signals to the clock is impaired.

The fact that the *elf3* mutant is more impaired in its response to blue light than in its response to red light further suggests that a blue-light photoreceptor plays a role in the photoperiodic control of flowering.

Genes That Control Flowering Time

As we have seen, the Maryland Mammoth mutation converts a day-neutral variety of tobacco into an SDP. Analysis of this flowering-time mutant by W. Garner and H. Allard led to the discovery that plants can measure photoperiod as an environmental cue for floral induction, and it identified the Maryland Mammoth gene as one of the genes that regulates this process. The most extensive studies of flowering-time genes have been performed in pea (*Pisum sativum*) and *Arabidopsis thaliana*. In these species a range of mutations can alter various aspects of the photoperiodic control of flowering.

For example, mutations in certain genes convert these LDPs into day-neutral plants that can flower rapidly. Mutations in other genes result in day-neutral plants that exhibit a long delay in flowering. Because these mutations eliminate the response to photoperiod, the normal (wild-type) products of these genes are thought to be involved in photoperiodism. Other mutations can delay or promote flowering without affecting the photoperiodic response. These mutants are thought to affect genes involved in the *autonomous* regulation of flowering.

Since most biological regulatory systems are under both positive and negative control, it is not surprising to find mutations that promote or delay flowering. Inasmuch as most mutations result in a loss or reduction of gene activity, the wild-type role of genes that delay flowering when mutated is to promote flowering.

Similarly, mutations that promote flowering are likely to occur in genes that normally inhibit the process. These genes are promoters and inhibitors of flowering in genetic terms, but the mutant phenotype alone does not usually reveal the flowering mechanism that is affected. For example, a mutation that eliminates the response to photoperiod could either block the production of a flowering hormone (floral stimulus) or, alternatively, interfere with the ability of the target cells in the meristem to receive the hormonal signal (meristem competence).

Arabidopsis plants containing mutations that promote or delay flowering have been isolated. The wild type and a *gigantea* (*gi*) mutant that were grown in inductive photoperiods (long days). The plants are shown at the stage when flower buds have become visible. Flowering in the *gigantea* mutant is delayed; thus this plant has been growing approximately three times as long as the wild type and has formed many more leaves from the primary apical meristem. Because a mutation in the *GIGANTEA* gene delays flowering, the normal role of this gene is to promote flowering. The wild type and an *elf3* mutant plant that were grown in noninductive photoperiods (short days). The mutant phenotype indicates that the normal role of the *ELF3* gene is to repress flowering.

One way to explore the role of these genes is through grafting studies. For example, if the apical bud from a mutant plant with a delayed-flowering phenotype flowers normally when grafted onto a wild-type shoot, the mutant shoot apex is clearly competent to receive and respond to flowering signals. One might then suspect that the mutation affects the production of flowering signals by leaves. If this were the case, a wild-type shoot apex region should exhibit delayed flowering when grafted onto a mutant shoot. Grafting studies of this sort have identified two types of genes in pea: those that control the production of transmissible promoters and inhibitors of flowering, and those that control the sensitivity of the shoot apical meristem to these signals. For example, the *GIGAS* gene is thought to be involved in the production of a transmissible flowering promoter. The wild-type pea exhibits a quantitative LDP phenotype. Plants homozygous for the recessive *gigas* mutation exhibit severely delayed flowering. For example, when grown under a 16-hour photoperiod, the wild type flowers after producing 16 leaves, whereas the *gigas* mutant flowers after producing 43 leaves. Delayed flowering in *gigas* plants could be caused either by the presence of inhibitors or by the absence of a floral stimulus. To test for inhibitors, researchers grafted the shoot of an early flowering day-neutral line onto *gigas* stock. The time of flowering of the scion was unaffected by grafting to the mutant, indicating that the mutant does not produce flowering inhibitors. However, when a *gigas* shoot was grafted onto either the early-flowering or the wild-type stock, flowering was greatly accelerated. Instead of flowering after producing 40 leaves or more, the plants flowered after producing only 14 or 23 leaves, respectively.

Because the *gigas* mutant shoot apex flowers readily after grafting, its ability to respond to the flowering stimulus from the leaves is not affected by the mutation. The mutation is therefore thought to affect the production of a floral stimulus. Another approach to elucidating the floral induction process is to isolate the genes identified by mutations and to study their properties. Once the gene is cloned, the encoded protein can be overexpressed in *E. coli* and purified in sufficient quantities to study its biochemical properties and to produce specific antibodies. The antibodies can be used to determine the tissues in which the protein is produced, as well as its location within the cell. The localization may provide clues to the function of the protein. For example, a protein in the shoot apex may be involved in the reception of flowering signals, whereas a protein present in the leaves may be involved in the production of flowering signals.

The *constans* mutants of *Arabidopsis* are insensitive to day length and flower later than the wild type under long-day conditions. The *CONSTANS* gene of *Arabidopsis* has been cloned and shown to encode a protein that shows similarities to zinc finger transcription factors, suggesting that it acts as a transcriptional regulator. The wild-type *CONSTANS* gene is expressed in leaves, and the level of *CONSTANS* mRNA increases during inductive photoperiods.

This pattern of expression, along with the photoperiod-insensitive mutant phenotype, implicate this gene in the photoperiodic control of flowering.

The Different Effects of Two Different Gibberellins on Flowering (Spike Length) and Elongation (Stem Length)

In the long-day plant *Lolium temulentum*, GA_{32} strongly promotes flowering, but has only a small effect on stem elongation. In contrast, GA_1 strongly promotes stem growth but has only a small effect on flowering. Gibberellin-treated plants were maintained in short days and, for comparison, the response to a single 24-hour day is shown. This long day strongly promotes flowering but has almost no effect on stem length.

The Influence of Cytokinins and Polyamines on Flowering

Sometimes changes normally associated with flowering can be caused by treatments that do not themselves bring about floral initiation. For example, one of the earliest events observed at the shoot apex following photoperiodic induction is a transitory increase in the number of cells undergoing mtosis.

When cytokinins are applied to the stem apex of the long-day plant white mustard (*Sinapis alba*), they cause an increase in mitotic activity similar to that caused by exposure to a single long day. However, whereas the long day induces flowering, the cytokinin treatment does not. Cytokinin may therefore be a component of the floral stimulus, although it is insufficient by itself to bring about flowering.

Other potential components include the polyamines. For example, photoperiodic induction of the LDP *Sinapis alba* was correlated with a large increase in the concentration of putrescine, the major polyamine, in the phloem exudate from the leaves. Spraying the leaves with an inhibitor of putrescine biosynthesis lowered the level of putrescine in the phloem exudate and also inhibited flowering. This result suggests that putrescine may be a component of the floral stimulus in *S. alba*.

In addition to receiving signals from the leaves, the shoot apex may accumulate flowering signals from the roots. Classic experiments with cuttings have shown that the initiation and growth of adventitious roots inhibit flowering in the shoot. In many species, root removal promotes flowering. Thus, at least in certain species, the root appears to produce one or more flowering inhibitors. In other cases, the roots may contribute floral activators.

For example, both GA and cytokinin are synthesized in the root and transported to the shoot via the xylem. During photoperiodic induction in *Sinapis alba,* the amount of cytokinin present in the xylem exudate from the root increases considerably. This finding indicates that signals emanating from the leaf are transported both to the shoot apex and to the root during photoperiodic induction. In response to the signal from the leaves, the roots produce and export additional chemical signals that modulate the flowering response.

The Contrasting Effects of Phytochromes A and B on Flowering

Phytochrome is encoded by members of a multigene family, and it is possible that the effects of red and far-red light described in the previous section are mediated by different types of phytochrome (A and B). In LDPs such as *Arabidopsis*, pea (*Pisum sativum*), and sorghum (*Sorghum bicolor*), phytochrome B (phyB) appears to be an *inhibitor* of flowering, since mutations in the gene *PHYB* that eliminate or reduce the amount of the phyB protein cause more rapid flowering.

The promotion of flowering by far-red light may result from a decrease in the amount of the Pfr form of phyB, which would have the same effect on the phyB Pfr levels as the mutations that reduce the total amount of phyB. As already discussed, red-light interruptions of a dark period prevent flowering in SDPs; therefore, the Pfr form of phyB may be an *inhibitor* of flowering in SDPs as well.

In the LDPs *Arabidopsis* and pea, mutations in the *PHYA* gene interfere with the promotion of flowering by long days; thus the Pfr form of phyA may promote flowering in LDPs. The effect of the *phyA* mutation is quite strong in pea (the mutant plants are essentially unable to respond to inductive photoperiods) whereas the *phyA* mutation in *Arabidopsis* has only a slight effect on the ability to respond to inductive photoperiods.

The species difference in the effect of *phyA* mutations is probably due to the fact that pea relies entirely on phyA to sense inductive photoperiods, whereas *Arabidopsis* also uses a blue light photoreceptor for this purpose.

A Gene that Regulates the Floral Stimulus in Maize

Maize plants flower after producing a fixed number of leaves regardless of photoperiod—a characteristic of autonomous regulation. If the shoot apical meristem from a maize plant is excised, and the "explant" is cultured in vitro, it is possible to regenerate a complete plant that will eventually flower. As long as the explant includes 4-6 leaf primordia, the meristem "remembers" its developmental stage at the time of excision and produces only the additional number of leaves needed to make up the fixed amount before flowering. However, if the explant contains fewer than 4 leaf primordia, the apex "forgets" its developmental stage and produces a complete set of new leaves before flowering.

The *indeterminate 1* (*id1*) mutant of maize is unable to undergo a normal transition to flowering, and produces many more leaves than the wild type. Moreover, the flowers produced by *id1* mutants are abnormal, having vegetative characteristics. The *ID1* gene encodes a zinc-finger protein, suggesting that it functions as a transcriptional regulator. The expression pattern of *ID1* differs from that of many other flowering time control genes isolated from *Arabidopsis* in that it is expressed in young leaves rather than in the shoot apical meristem.

Are the effects of the *ID1* gene "cell autonomous", that is, restricted to those cells in which the gene is expressed, or does the *ID1* gene regulate the production of a transmissible signal that can affect cells beyond the site of gene expression? Traditionally, grafting experiments between the mutant and the wild type have been carried out to answer such questions, but maize (a monocot) lacks a vascular cambium and is therefore not suitable for grafting experiments.

In a second set of experiments, Colasanti et al. (1998) did the next best thing to grafting by analysing plants that were genetic chimeras (that is, containing cells with different genotypes) for *ID1* function. The mutant allele of *id1* analysed was caused by the insertion of a transposable element, and the spontaneous excision of this element during development generated clonal sectors of wild-type *ID1*-expressing cells against an overall mutant *id1* background.

It was found that chimeric *id1* plants containing sectors of restored *ID1* function flowered earlier than completely mutant *id1* plants, even though the shoot apical meristems of these chimeric plants remained mutant. This experiment is comparable to grafting a wild type scion onto an *id1* mutant stock, and finding that the *id1* stock flowered earlier as a result. This observation in combination with the expression data provides compelling evidence for the involvement of *idt1* in the production in leaves of a transmissable signal from immature leaves that induces flowering in the apex.

The *Arabidopsis* genome contains several potential homologs of the maize *ID1* gene, but no *id1*-like mutants have been identified with a phenotype similar to that in maize. All of the genes involved in the autonomous pathway of *Arabidopsis* thus far seem to be expressed exclusively in the apical meristem.

Chapter

8

Secondary Metabolites and Plant Defence

Structure of Various Triterpenes

Digitoxigenin is the aglycone (sugarless) triterpene portion of the naturally occurring digitanides, which contain one molecule of the rare sugar D-digitoxose and one molecule of acetyldigitoxose. The yam *Dioscorea* contains saponins—e.g., yamogenin—that are widely used as starting materials in the synthesis of progesterone-like compounds for birth control pills.

The Shikimic Acid Pathway

Plant phenolics are biosynthesized by several different routes and thus constitute a heterogeneous group from a metabolic point of view. Two basic pathways are involved: the shikimic acid pathway and the malonic acid pathway. The shikimic acid pathway participates in the biosynthesis of most plant phenolics. The malonic acid pathway, although an important source of phenolic secondary products in fungi and bacteria, is of less significance in higher plants.

The shikimic acid pathway converts simple carbohydrate precursors derived from glycolysis and the pentose phosphate pathway to the aromatic amino acids. One of the pathway intermediates is shikimic acid, which has given its name to this whole sequence of reactions.

The well-known, broad-spectrum herbicide glyphosate (available commercially as Roundup) kills plants by blocking a step in this pathway. The shikimic acid pathway is present in plants, fungi, and bacteria but is not found in animals. Animals have no way to synthesize the three aromatic amino acids—phenylalanine, tyrosine, and tryptophan—which are therefore essential nutrients in animal diets.

Cell Walls: Structure, Biogenesis and Expansion

Terminology for Polysaccharide Chemistry

Polysaccharides are named after the principal sugars they contain. For example, a *glucan* is a polymer made up of glucose; a *xylan* is a polymer made up of xylose; a *galactan* is a polymer made up of galactose.

For branched polysaccharides, the backbone of the polysaccharide is usually given by the last part of the name. Thus, *xyloglucan* has a glucan backbone (a linear chain of glucose residues) with xylose sugars attached to it in the side chains; *arabinoxylan* has a xylan backbone (made up of xylose subunits) with arabinose side chains. However, a compound name does not necessarily imply a branched structure.

For example, "glucomannan" is the name given to a polymer containing both glucose and mannose in its backbone. Because sugars may assume multiple forms and may be connected in many ways, the nomenclature for sugars is not simple and is briefly summarized here.

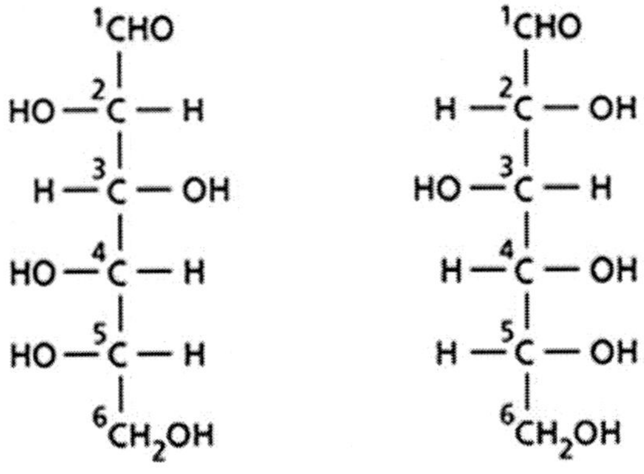

L-Glucose D-Glucose

Figure: *D-and L-glucose shown in the linear form. The carbons are numbered from the end with the aldehyde (CHO) (C = O) group.*

Monosaccharides (single-sugar residues) are polyhydroxyl aldehydes or ketones with the empirical formula $(CH_2O)_n$. They typically contain one asymmetric carbon, so they can exist as two stereoisomers, designated as the *D* or *L* enantiomers.

Monosaccharides usually form a cyclic, rather than a linear, structure, because the carbonyl oxygen can react with the hydroxyl groups, typically forming either five-member rings (furanose rings) or six-member rings (pyranose rings).

These rings have a new chiral centre and may take two forms, designated as α or β anomers. Thus, β-*D*-glucopyranose is *D*-glucose taking the form of a six-member ring in the β configuration. In solution, monosaccharides freely mutorotate (convert between α or β anomers); when linked in polymers, the anomeric configuration is stable.

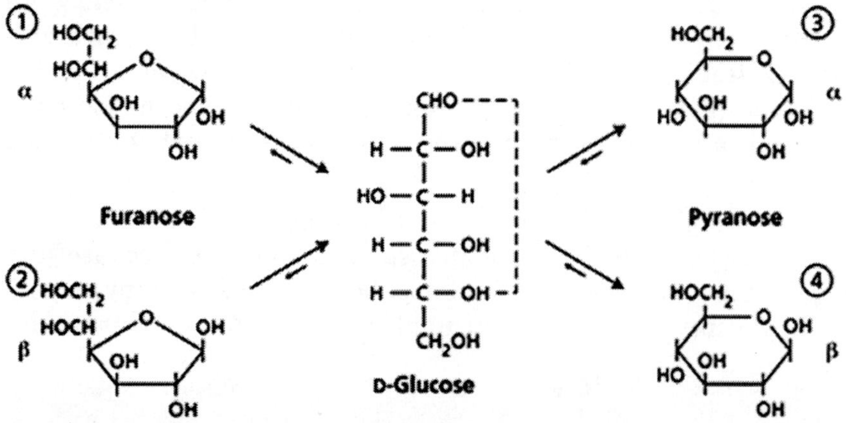

Figure: *Glucose may form five-or six-member rings, in the a or b configuration. The five-member rings (furanose; structures 1 and 2) are unstable and make up less than 1% of the total. The six-member rings (pyranose) are found mostly in the b configuration (64% of the total); the remainder (36%) are found in the a form.*

The sugars in polysaccharides are linked together by O-glycosidic bonds. We indicate the linkage pattern by specifying the carbon atoms connected by the glycosidic bond. In cellulose, for example, carbon atom 1 is linked to carbon atom 4 in the neighbouring glucose residues.

Thus, cellulose is designated as a (1→4) β-*D*-glucan, because it is a string of β-*D*-glucose molecules linked through carbons 1 and 4. It may also be called a polymer of (1→4)-linked β-*D*-glucopyranose.

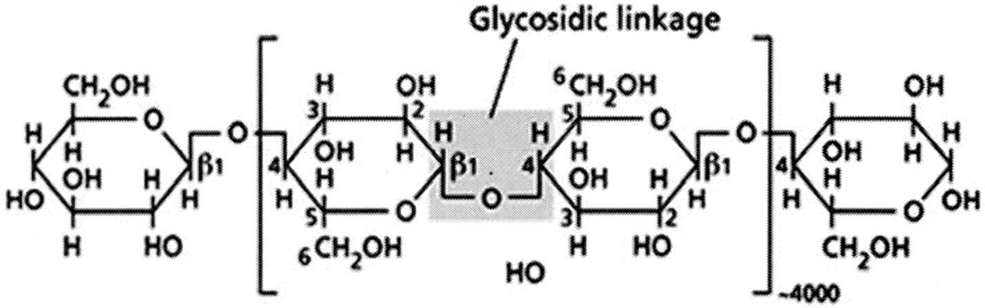

Figure: *A linear chain of 1→4-linked b-D-glucopyranose, showing the numbering system for the carbons and the configuration of the glycosidic bond. This is the glucan that makes up cellulose. The sugar rings are connected through an oxygen atom (the glycosidic oxygen). Because of the configuration of the glycosidic bonds, the repeating unit in this polymer is cellobiose, a disaccharide.*

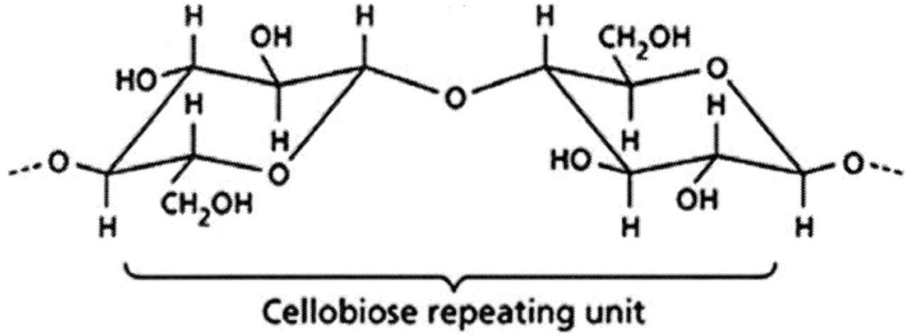

Figure: *The repeating cellobiose unit, with the glucose residues drawn in the conformational ("chair") form. This way of drawing the glucose ring represents better the actual positions of the individual atoms than in the Haworth project.*

Molecular Model for the Synthesis of Cellulose and Other Wall Polysaccharides That Consist of a Disaccharide Repeat

Cellulose synthase contains two binding sites for UDP-glucose (or other sugar nucleotides). After binding UDP-glucose (step 2), the sugars are transferred to the nonreducing end of the glucan (step 3). At step 4, the glucan chain is translocated by two glucose residues, restoring the complex for the start of another round of synthesis (step 1). These genes code for glycosyl transferases with two catalytic sites for transferring glycosyl residues from their nucleotide donors to the nascent polymer acceptor. This two-site arrangement may explain why the cellobiose disaccharide is the repeating subunit in cellulose:

It has been proposed that two glucose residues are added to the glucan chain for each catalytic cycle of the enzyme. A similar arrangement in other polysaccharide synthases could explain why disaccharide repeats commonly make up the backbones of many other cell wall polysaccharides in plants.

The Mechanical Properties of Cell Walls: Studies With *Nitella*

The small size of typical plant cells (20 to 100 μm) has been a serious impediment to the study of cell wall mechanical properties. To measure the extensibility of isolated cell walls from higher plants, researchers must place entire frozen and thawed organs or tissues in an *extensometer*. Although such measurements can give qualitative answers, the extensions measured cannot be related to the growth of living cells in a quantitative way, because the applied stress is uniaxial (unidirectional), whereas the in vivo stress due to turgor pressure is multiaxial (multidirectional).

The giant internodal cells of the freshwater green algae *Nitella* and *Chara* have figured prominently in classic studies on ion transport, water permeability, and cytoplasmic streaming (Hope and Walker 1975). These algae are characterized by erect, branched coenocytic (multinucleate) filaments, differentiated into nodes and internodes, each node bearing a whorl of specialized lateral internodes.

The internode cell begins as a shortened disc 20 mm long, which elongates to produce a slender cylinder about 6 cm long and 0.5 mm wide at maturity. The cell expands by diffuse growth in a predominantly longitudinal direction. Because individual internode cells are large, their cell walls can be isolated and the mechanical properties of the walls measured. It is even possible to study the anisotropic (unequal) mechanical properties of individual cell walls by the use of multiaxial stress to mimic the in vivo turgor pressure.

The experimental setup for studying multiaxial extensibility in *Nitella* internode cells. Empty cell wall cylinders are glued to a pipette and inflated with mercury under pressure. Liquid mercury is insoluble and therefore inert under these conditions. In response to the multiaxial stress caused by the pressurized mercury, the cell wall cylinder expands, and investigators can monitor the expansion in both length and width by viewing the process through a horizontal microscope (Metraux et al. 1980).

With the multiaxial stress procedure on *Nitella* internodes, one can use microtubule inhibitors to alter the orientation of the cellulose microfibrils and test the hypothesis that microfibril orientation is responsible for the mechanical anisotropy of the cell wall. The multiaxial extension of control cells is highly anisotropic. Not only is extension in length four times greater than extension in diameter, but much of the extension in length is irreversible, whereas the extension in diameter is almost completely reversible.

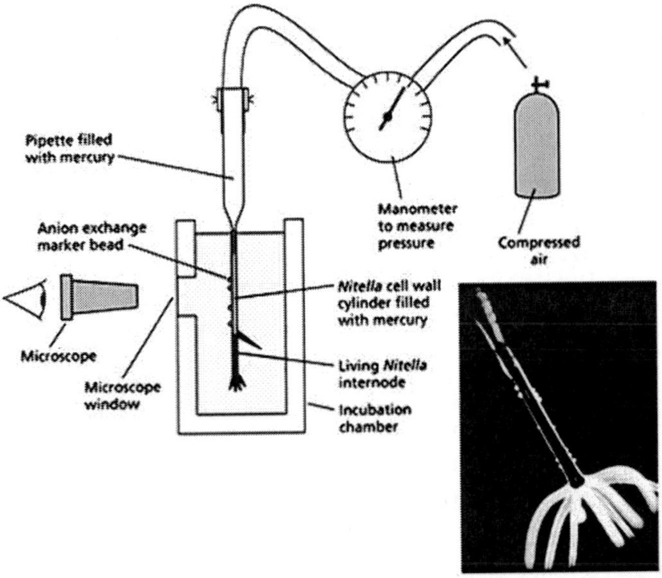

Figure: *Apparatus for measuring the multiaxial extensibility of Nitella cell walls. The pressure source is a compressed air tank. Air passes through high-pressure tubing into a manometer and from there into a mercury-filled pipette glued to the open end of a Nitella cell wall cylinder. A living internode is maintained at the distal end. The wall is immersed in a perfusion chamber with a glass window. Displacement of resin beads attached to the surface of the wall is measured with a horizontally oriented microscope. The inset shows a view of a mercury-inflated cell wall cylinder with marker beads attached as seen through the traveling microscope.*

These results are consistent with the anisotropic growth pattern of the cells. The effects of pretreating the cells with the microtubule poison isopropyl *N*-phenylcarbamate (IPC), which, like colchicine, induces the random deposition of cellulose microfibrils in the wall. The result is that the extensions of the wall in length and diameter are about equal, consistent with the isotropic growth pattern induced by this inhibitor.

Nitella has also proven to be useful for studying cell wall properties in vivo. Many of the concepts applied to contemporary studies on the biophysical control of cell expansion are derived from the pioneering experiments on living *Nitella* cells by Paul Green and his colleagues at Stanford University (Green 1968). An important insight to come out of such studies is that the inner 25% of the cell wall determines the polarity of cell expansion.

Taking advantage of the fact that cellulose microfibrils are crystalline and therefore exhibit birefringence (double refraction) when viewed with polarized light, Paul Richmond at the University of the Pacific in California devised an optical technique to measure cell expansion in length and diameter, and the orientation of the cellulose microfibril in the cell wall, simultaneously (Richmond 1983). This technique made it possible to determine the relationship between microfibril angle and the direction of cell expansion.

The microfibril angle began to change from transverse to random almost immediately after colchicine was added to the growth medium. However, the anisotropy of growth did not change until about 8 hours later. The transition from anisotropic growth to isotropic growth is complete after a relative increase in cell surface area of about 25%.

Since wall thickness remains constant during this period because of continued wall synthesis, we can infer that the directionality of cell expansion changes when the inner 25% of the cell wall has been replaced. Thus the inner 25% of the cell wall appears to represent the load-bearing portion of the wall. By implication, the outer 75% has become too weakened by stretching to restrict cell expansion.

Growth and Development: The Analysis of Plant Growth

As we have seen, new cells form continually in the apical meristems. Cells enlarge slowly in the apical meristem and more rapidly in the subapical regions. The resulting increase in cell volume can range from severalfold to one hundredfold, depending on the species and environmental conditions. Classically, plant growth has been analysed in terms of cell number or overall size (i.e., mass). However, these measures tell only part of the story.

Tissue growth is neither uniform nor random. The derivatives of the apical meristems expand in predictable and site-specific ways, and the expansion patterns in these subapical regions largely determine the size and shape of the primary plant body. The total growth of the plant can be thought of as the sum of the local patterns of cell expansion.

The analysis of the motions of cells or "tissue elements" (and the related problem of cell expansion) is called *kinematics*. In this section we will discuss both the classical definitions of growth and the more modern, kinematic approach. As we will see, the advantage of the kinematic approach is that it allows one to describe the growth patterns of organs mathematically in terms of the expansion patterns of their component cells.

Plant Growth Can Be Measured in Different Ways

Growth in plants is defined as an irreversible increase in volume. The largest component of plant growth is cell expansion driven by turgor pressure. During this process, cells increase in volume manyfold and become highly vacuolate. However, size is only one criterion that may be used to measure growth.

Growth also can be measured in terms of change in fresh weight—that is, the weight of the living tissue—over a particular period of time. However, the fresh weight of plants growing in soil fluctuates in response to changes in water status, so this criterion may be a poor indicator of actual growth. In these situations, measurements of dry weight are often more appropriate.

Cell number is a common and convenient parameter by which to measure the growth of unicellular organisms, such as the green alga *Chlamydomonas* (Figure). In multicellular plants, however, cell number can be a misleading growth measurement because cells can divide without increasing in volume.

For example, during the early stages of embryogenesis, the zygote subdivides into progressively smaller cells with no net increase in the size of the embryo. Only after it reaches the eight-cell stage does the increase in volume begin to mirror the increase in cell number. Because the zygote is an especially large cell, this lack of correspondence between an increase in cell number and growth may be unusual, but it points out the potential problem in equating an increase in cell number with growth.

Although cell number may not always be a reliable measure of plant growth, under most circumstances dividing cells, particularly in meristems, double in volume during their cell cycle. Therefore, an increase in cell number, such as the increase brought about by the activity of the apical meristems, does contribute to plant growth. However, the largest component of plant growth is the rapid cell expansion that occurs in the subapical region after cell division ceases.

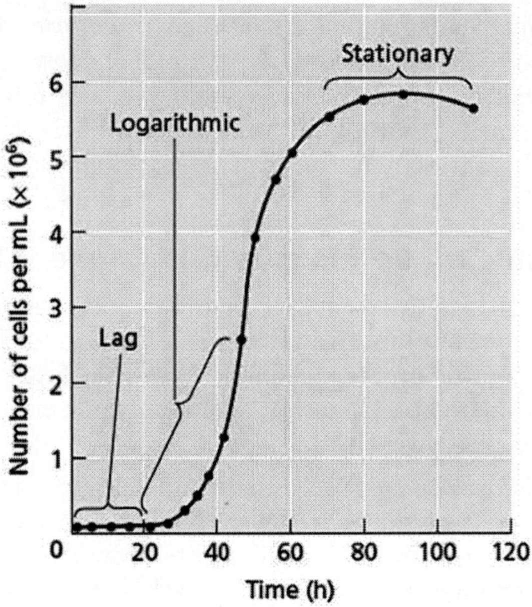

Figure: *Growth of the unicellular green alga Chlamydomonas. Growth is assessed by a count of the number of cells per milliliter at increasing times after the cells are placed in fresh growth medium. Temperature, light, and nutrients provided are optimal for growth. An initial lag period during which cells may synthesize enzymes required for rapid growth is followed by a period in which cell number increases exponentially. This period of rapid growth is followed by a period of slowing growth in which the cell number increases linearly. Then comes the stationary phase, in which the cell number remains constant or even declines as nutrients are exhausted from the medium.*

Because all the cells of the plant axis elongate under normal conditions, the greater the number of cells produced by the apical meristem, the longer the axis will be. For example, when Arabidopsis plants are transformed with a gene that encodes cyclin, a key component of the cell cycle regulatory machinery, the cells of the apical meristem progress through their cell cycles more rapidly, so more cells form per unit time. As a result, the roots of these transgenic plants have more cells and are substantially longer than the roots of wild-type plants grown under similar conditions. New cells form continually in the apical meristems. With each new round of cell division and associated cell expansion, the older derivatives are displaced a small distance from the apex. As the cells recede farther from the apex, the rate of displacement is greatly accelerated. By viewing plant growth as a process of cell displacement from the apex, we can apply the principles of kinematics.

The Production of Cells by the Meristem Is Comparable to a Fountain

Moving fluids such as waterfalls, fountains, and the wakes of boats can generate specific forms. The study of the motion of fluid particles and the shape changes that the fluids undergo is called kinematics. The ideas and numerical methods used to study these fluid forms are useful for characterizing meristematic growth. In both cases, an unchanging form is produced even though it is composed of moving and changing elements. An example of an unchanging form composed of changing and displaced elements in plants is the hypocotyl hook of a dicot, such as the common bean (Figure above). As the bean seedling emerges from the seed coat, the apical end of the hypocotyl bends back on itself to form a hook. The hook is thought to protect the seedling apex from damage during growth through the soil. During seedling growth (in soil or dim light) the hook migrates up the stem, from the hypocotyl into the epicotyl and then to the first and second internodes, but the form of the hook remains constant.

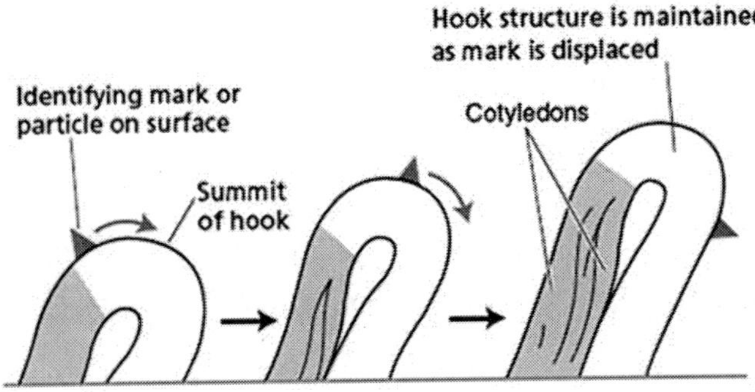

Figure: *The dicot hypocotyl hook is an example of a constant form composed of changing elements. The hooked form is maintained over time, while different tissues first curve and then straighten as they are displaced from the seedling apex during growth. If a mark is placed at a fixed point on the surface, it will be displaced (indicated by the arrow), appearing to flow through the hook over time.*

If we mark a specific epidermal cell on the seedling stem located close to the seedling apex, we can watch it as it flows into the hook summit, then down into the straight region below the hook. The mark is not crawling over the plant surface, of course; plant cells are cemented together and do not experience much relative motion during development. The change in position of the mark relative to the hook implies that the hook is composed of a procession of tissue elements, each of which first curves and then straightens as it is displaced from the plant apex during growth. The steady form is produced by a parade of changing cells.

A root tip is another example of a steady form composed of changing tissue elements. Here, too, the form is observed to be steady only when distance is measured from the root tip. A region of cell division occupies perhaps 2 mm of the root tip.

The elongation zone extends for about 10 mm behind the root tip. Phloem differentiation is first observed beginning at 3 mm from the tip, and functional xylem elements may be seen at about 12 mm from the tip. A marked cell near the tip will seem to flow first through the region of cell division, then through the elongation zone and into the region of xylem differentiation, and so on. This shifting implies that developing tissue elements first divide and elongate, and then differentiate.

In an analogous fashion, the shoot bears a succession of leaves of different developmental stages. During a period of 24 hours, a leaf may grow to the same size, shape, and biochemical composition that its neighbor had a day earlier. Thus, shoot form is also produced by a parade of changing elements that can be analysed with kinematics. Such an analysis is not merely descriptive; it permits calculations of the growth and biosynthetic rates of individual tissue elements (cells) within a dynamic structure.

Tissue Elements Are Displaced During Expansion

As we have seen, growth in shoots and roots is localized in regions at the tips of these organs. Regions with expanding tissue are called growth zones. With time, meristems move away from the plant base by the growth of the cells in the growth zone.

If successive marks are placed on the stem or root, the distance between the marks will change, depending on where they are within the growth zone. In addition, all of these marks will move away from the tip of the root or shoot, but their rate of movement will differ depending on their distance from the tip.

From another perspective, if you were to stand at the tip of a root that had marks placed at intervals along the axis, you would see that all marks would move farther away from you with time. The reason is that discrete regions on the plant axis experience displacement as well as expansion during growth and development.

As Regions Move Away from the Apex, Their Growth Rate Increases

As a given region of the plant axis moves away from the apex, its growth velocity increases (the rate of elongation accelerates) until a constant limiting velocity is reached equal to the overall organ extension rate. The reason for this increase in growth velocity is that with time, progressively more tissue is located between the moving particle and the apex, and progressively more cells are expanding, so the particle is displaced more and more rapidly. In a rapidly growing maize root, a tissue element takes about 8 hours to move from 2 mm (the end of the meristematic zone) to 12 mm (the end of the elongation zone).

Beyond the growth zone, elements do not separate; neighbouring elements have the same velocity (expressed as the change in distance from the tip per unit of time), and the rate at which particles are displaced from the tip is the same as the rate at which the tip moves through the soil. The root tip of maize is pushed through the soil at 3 mm h^{-1}. This is also the rate at which the nongrowing region recedes from the apex, and it is equal to the final slope of the growth trajectory.

The Growth Velocity Profile Is a Spatial Description of Growth

The velocities of different tissue elements are plotted against their distance from the apex to give the spatial pattern of growth velocity, or growth velocity profile. Growth velocity increases with position in the growth zone. A constant value is obtained at the base of the growth zone. The final growth velocity is the final, constant slope of the growth trajectory equal to the elongation rate of the organ, as discussed in the previous section. In the rapidly growing maize root, the growth velocity is 1 mm h^{-1} at 4 mm, and it reaches its final value of nearly 3 mm h^{-1} at 12 mm.

If the growth velocity is known, the relative elemental growth rate, which represents the fractional change in length per unit time, can be calculated. The relative elemental growth rate shows the location and magnitude of the extension rate and can be used to quantify the effects of environmental variation on the growth pattern.

The Relative Elemental Growth Rate

One way to characterize the growth pattern along an axis mathematically is to plot the distance of a tissue element from the apex versus time. The resulting curve is called the growth trajectory.

The slope of the curve at any given point is equivalent to the growth velocity at that region of the axis. With the apex as the reference point, the growth trajectory is concave. Marked tissue elements at different locations may be followed simultaneously to generate a family of growth trajectories.

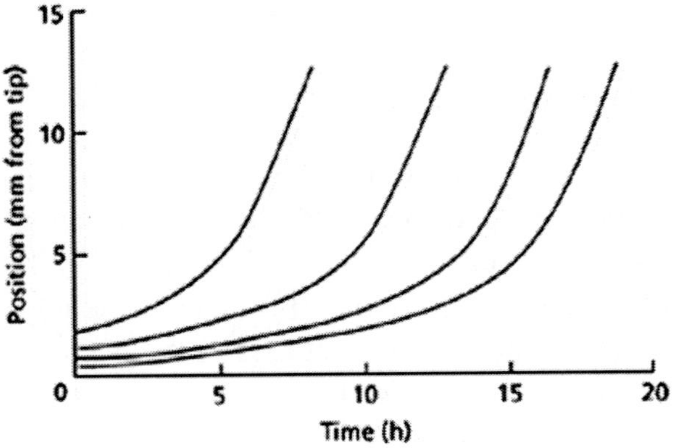

Figure: *Growth trajectories of* Zea mays *(maize):Growth trajectories plot the positions of points spaced along the root axis as a function of time. Note that distance is measured from the root tip. This is termed a material specification of growth because real or material particles are followed.*

As the regions of the plant axis move away from the apex, their growth velocity increases (i.e., the cells accelerate) until a constant limiting velocity is reached equal to the overall organ extension rate. The reason for this increase in growth velocity is that with time, progressively more tissue is located between the moving particle and the apex, and progressively more cells are expanding, so the particle is displaced more and more rapidly. In a rapidly growing maize root, a tissue element takes about 8 hours to move from 2 mm (the end of the meristem) to 12 mm (the end of the elongation zone). The root tissue element accelerates through this region.

Beyond the growth zone, elements do not separate; neighbouring elements have the same velocity, and the rate at which particles are displaced from the tip is the same as the rate at which the tip is pushed away from the soil surface. The root tip of maize is pushed through the soil at 3 mm h^{-1}. This is also the rate at which the nongrowing region recedes from the apex, and it is equal to the final slope of the growth trajectory.

The growth trajectory is a material description of growth because an individual element is followed through time. The growth trajectory provides a way to convert spatial information into a developmental time course. For example, if the amount of suberin in the root increases with position, the growth trajectory will give information on the time required for the observed accumulation of suberin in cells moving between any two spatial points.

To characterize the local expansion rate, we must consider the growth velocities at the apical and basal ends of a small segment of tissue. If the apical end of the segment is moving faster than the basal end, the segment is elongating. The velocity difference, divided by the segment length, gives the relative growth rate of the segment. Now, imagine that the segment shrinks to a point x. To find the relative growth rate, r, at point x, we can use calculus: The value of r is given by the velocity gradient, dv/dx, where v is the growth velocity. This measure of the local growth rate is called the relative elemental growth rate (Erickson and Sax 1956). It represents the fractional change in length per unit time and has units of h^{-1}. If the growth velocity is known, the relative elemental growth rate can be evaluated by differentiation of the growth velocity with respect to position. Or we can approximate r by measuring the relative growth rates of closely spaced segments of initial length L. For each segment, $r = (1/L)(dL/dt)$.

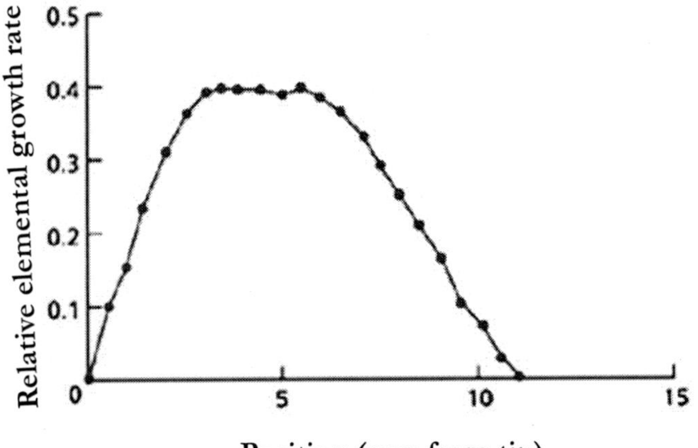

***Figure:** Relative elemental growth rate of* Zea mays *(maize): The relative elemental growth rate tells us the rate of expansion of any particular point on the root. It is the most useful measure of root growth rate for the physiologist because it tells where the most rapidly expanding regions are located.*

The relative elemental growth rate profile, $r(x)$, is a spatial description of the expansion pattern in the growth zone. In the maize root, the relative elemental growth rate has a maximum of 0.40 h^{-1} at 4 mm and drops to zero by 12 mm. That is, the tissue element at 4 mm is expanding 40% h^{-1}, while tissue beyond 12 mm has ceased expansion. Again, to understand the profile in physiological terms, we must consider that individual cells expand more and more rapidly as they are displaced to 4 mm and then less and less rapidly as they are displaced through the basal part of the growth zone.

The relative elemental growth rate profile shows the location and magnitude of the extension rate and can be used to quantify the effects of environmental variation on the growth pattern. It is also the basis for studying the physiology of growth. The graph of $r(x)$ can be compared to the distribution pattern of any substance thought to be regulating the growth rate. A close correlation between the two curves would provide circumstantial evidence for the role of the substance as a limiting factor for growth.

Embryonic Dormancy

Embryo Maturation Requires Specific Gene Expression

The Arabidopsis embryo enters dormancy after it has generated about 20,000 cells. Dormancy is brought about by the loss of water and a general shutting down of gene transcription and protein synthesis—not only in the embryo, but also throughout the seed. To adapt the cell to the special conditions of dormancy, specific gene expression is required. For example, the *A*bscisic *A*cid *I*nsensitive*3* (*ABI3*) and *FUSCA3* genes are necessary for the initiation of dormancy and are sensitive to the hormone abscisic acid, which is the signalling molecule that initiates seed and embryo dormancy. *ABI3* also controls the expression of genes encoding the storage proteins that are deposited in the cotyledons during the maturation phase of embryogenesis.

The *LE*AFY *C*OTYLEDON*1* (*LEC1*) gene also is active in late embryogenesis. Because *lec1* mutants cannot survive desiccation and do not enter dormancy, the embryos die unless they are rescued through isolation before desiccation occurs. The rescued embryos will germinate in culture and produce fertile plants, which are like wild-type plants except that they lack the 7S storage protein and they have leaflike cotyledons with trichomes on their upper surface.

The normal appearance and development of the mature *lec1* mutants indicates that the *LEC1* gene is required only during embryogenesis. Although the most obvious defects of the *lec1* mutants are seen only in the maturation phase embryo, mRNA from *LEC1* gene expression can be detected throughout embryogenesis. It has been proposed that *LEC1* is a general repressor of vegetative development and its expression is necessary throughout embryogenesis.

Polarity of Fucus *Zygotes*

Plant cells perceive their position within a tissue and assume the polarity appropriate for that tissue. In addition, cells can perceive and act on information coming from their environment to alter their polarity and redirect growth when necessary. Cellular polarity must be established to orient subsequent expansion. Cells that are initially apolar, such as zygotes, spores, and pollen grains, later become polarized and initiate tip growth and have been especially valuable in the analysis of the induction and establishment of polarity.

A wide variety of external gradients, including temperature, light intensity, pH, ions, and electric-potential gradients, can polarize and orient the axis of growth of these cells. Thus, polarity can be experimentally manipulated without the complexities associated with multicellular tissues. Much of our understanding of the establishment of polarity in tip-growing cells is derived from studies of rhizoid emergence in zygotes of the brown alga *Fucus*.

Polarity of the Fucus Zygote is Initially Unstable but Later Fixed

Unlike higher plants, and even unlike most other algae, certain brown algae, particularly *Fucus* and *Pelvetia*, shed their eggs and sperm into the sea, where fertilization occurs. The zygote initially lacks a cell wall and is completely apolar. However, a polar axis is established within a few hours of fertilization in response to unilateral light. The axis is not stable at first and can be changed many times simply by a change in the direction of the light.

The polar axis becomes fixed within 10 to 14 hours after fertilization, after which it cannot be altered by a change in the direction of the light, and a structure that will become the rhizoid begins to protrude from the darkened side.

Even though the zygote is still a single cell, it is polarized, with two structurally different poles, the thallus and rhizoid poles, which have radically different fates. The first zygotic division is asymmetric and occurs at right angles to the polar axis. It establishes two cells, each having the identity of the pole from which it emerges. The smaller *rhizoid cell* produces a filament of cells and the holdfast that attaches the plant to rocks. The larger *thallus cell* divides to produce a globular mass of cells, from which the main body of the plant, the thallus, forms.

Calcium Channel Redistribution Leads to a Transcellular Current

A dye known as fluorescent dihydropyridine (FL-DHP), which binds to calcium channels in mammalian cells and can interfere with calcium uptake, has been used as a probe to examine the polarization of the *Fucus* zygote cell membrane. During the first 6 to 8 hours after fertilization, and in response to unilateral light, FL-DHP receptors, presumably calcium channels, are redistributed and become localized on the dark side of the *Fucus* zygote at what will become the rhizoid pole. If the direction of the light is changed, the location of the FL-DHP–binding proteins also changes, again becoming concentrated on the dark side of the zygote. One of the earliest events in the establishment of polarity after exposure of the zygotes to unilateral light is the induction of a positive current that flows into the cell on the shaded side and loops back through the external medium. This transcellular current, which is due largely to calcium ions, is on the order of 100 pA (1 picoampere = 10^{-12} ampere) and develops in advance of any morphological sign of polarity. The relevance of this phenomenon to the establishment of cell polarity is indicated by the fact that the site at which positive charge flows into the cell is also the site of subsequent rhizoid outgrowth. Transcellular ionic currents also have been observed in some tip-growing cells of higher plants, including root hairs and germinating pollen grains, but they have not yet been detected in plant cells that expand by diffuse growth.

The ionic current in polarizing *Fucus* zygotes likely arises from the redistribution of ion channels in the plasma membrane. This current generates an electric-potential difference across the membrane that could alter other membrane properties, such as the distribution of membrane proteins. The current also alters the ionic composition of the cytosol; the magnitude of the effect depends on the ions that form the current, which vary in different cell types. Because the Ca^{2+} concentration in the cytosol is usually very low (between 10^{-6} and 10^{-8} μM), even a small increase in intracellular Ca^{2+} concentration could have a marked effect.

Changes in Cytosolic Calcium have been Measured

The fluorescent calcium-indicator dye *fura*-2 has been used to detect intracellular cytosolic calcium gradients in both plant and animal cells. Since the excitation spectrum of fura-2 shifts when it binds calcium, we can determine the intracellular concentrations of both forms of the dye (with and without bound calcium) by exciting them with the appropriate two wavelengths. The ratio of the two emissions provides a measure of the calcium concentration that is independent of the dye concentration. Using this technique to measure the cytosolic Ca^{2+} gradient in growing *Fucus* rhizoids, researchers observed a gradient from 450 nM at the tip to 100 nM at the base.

This calcium ion gradient may be important for the polarized assembly of actin filaments and vectoral discharge of Golgi-derived vesicles at the rhizoid pole. Calcium ion gradients also have been shown to be established in other tip-growing cells, such as germinating pollen grains, where the direction of growth and the region in which Golgi vesicles carrying cell wall components are deposited are predicted by the calcium gradient.

The Actin Cytoskeleton Becomes Polarized

Inhibition of protein synthesis by treatment with the drug *cycloheximide* delays rhizoid formation but does not stop the establishment of polarity. This result suggests that although rhizoid formation requires the synthesis of new proteins, the establishment of polarity does not, so polarity development must involve a redistribution of existing components. Actin filaments become preferentially localized in the rhizoid half of the zygote and extend from the vicinity of the nucleus to the plasma membrane at the site marked by localized FL-DHP–binding proteins as polarity develops.

Actin filaments are dynamic, and their assembly from a pool of monomeric actin subunits is highly regulated. Actin filaments are localized in the rhizoid half of the cell presumably in response to a change in the assembly or disassembly dynamics, and this localization involves a redistribution of the cellular actin.

The fixation of polarity can be blocked by *cytochalasin B*, a drug that destroys actin filaments. Treatment with cytochalasin B also disrupts the ionic current that develops during polarization. Actin filaments participate in cell secretion by transporting secretory vesicles to the plasma membrane in many types of cells.

Secretion of Golgi-Derived Vesicles Is Localized

Cytoplasmic organelles become asymmetrically distributed before the polarity of the cell is fixed. Golgi-derived vesicles accumulate at the site of the future outgrowth in the rhizoid half of the *Fucus* zygote as polarization is induced by light. For several hours after the induction of polarity, the asymmetric distribution of vesicles can be changed by subsequent treatment with light from a different direction.

During this phase when changes are possible, cytoplasmic vesicles accumulate at the prospective site of rhizoid formation on the shaded side of the cell. Subsequently, the vesicles fuse with the plasma membrane, secreting their contents into the cell wall, leading to rhizoid emergence. Evidence suggests that calcium gradients play a role in triggering vesicle discharge at the pole.

The fixation of light-induced polarity in *Fucus* zygotes also requires the presence of the cell wall. Removal of the wall with cell wall–digesting enzymes does not prevent the induction of polarity by unilateral light, but it does prevent fixation of the axis. The orientation of the embryonic axis remains unstable until a new cell wall is synthesized.

From these observations it is possible to envisage a chain of events that might occur during the establishment of cell polarity in tip-growing cells.

- The ionic currents that arise early in the process may initiate the formation of an intracellular Ca^{2+} gradient.
- The increase in intracellular calcium at the tip may help organize and stabilize the microfilament components of the cytoskeleton.
- The cytoskeleton directs vesicle traffic to the growing tip.
- A calcium gradient, perhaps immediately adjacent to the plasma membrane, may promote vesicle fusion with the plasma membrane, although this remains to be demonstrated.
- Finally, cell wall deposition fixes the polarity of the cell, possibly by interacting with transmembrane proteins.

The Preprophase Band of Microtubules

Microtubules in the cortical cytoplasm of growing plant cells are important in determining the orientation of cellulose microfibrils in the cell wall. When dividing cells enter the G_2 phase of the cell cycle, the interphase cortical microtubule array is replaced by a narrow band of interconnected microtubules and microfilaments encircling the nucleus. This circular array of microtubules and microfilaments is called the preprophase band. The preprophase band was first observed in 1966 by Jeremy Pickett-Heaps and Donald Northcote at

Cambridge University. It appears as a transient ring of cytoskeletal elements in the cortical cytoplasm, near the plasma membrane. Generally the preprophase band is 2 to 3 μm wide and contains from 10 to more than 100 microtubules, interspersed and interconnected with actin microfilaments.

Immunofluorescent staining of the cytoskeleton in onion (*Allium cepa*) root tip cells with an antibody to the microtubule protein, tubulin. As cells progress through their cell cycle, different cytoskeletal arrays that contain microtubules are formed. (A) A preprophase band appears after DNA replication is complete, but before the cell enters mitosis. It can be seen as two bright spots on either side of the unstained nucleus; the image of the band is fainter as it passes across the nucleus. (B) The mitotic spindle forms in late prophase, with spindle microtubules emanating from the polar regions to form the spindle fibres. (C) The spindle fibres hold the chromosomes (dark line) in the metaphase plate. In this cell the alignment of the plate is oblique, which is relatively unusual. (D) The chromosomes move toward their poles within the array of spindle microtubules. (E) The phragmoplast forms in telophase and is seen here as a bright band in the centre of the cell that consists of microtubules, microfilaments, and vesicles in the early stage of cell plate formation. (F) Phragmoplast microtubules form an annular ring around the perimeter of the cell plate and appear only at the edges of the cell. Bar = 10 μm. (Courtesy of A. Hardham.)

As the cell enters mitosis, microtubules begin to assemble at the surface of the nuclear envelope, and as the nuclear envelope breaks down, they invade the nuclear area to form the mitotic spindle. The preprophase band disappears as the spindle forms, leaving behind an actin-depleted zone. The plane in which cytokinesis occurs is determined before the cell initiates mitosis and is accurately predicted by the position of the preprophase band. It has been suggested that the preprophase band determines the orientation of the mitotic spindle and that it marks the site at which the new cell wall will form during cytokinesis, but this has not been established. The position of the preprophase band and the plane of cytokinesis both may be determined by another mechanism, possibly sensitive positional information.

Genes Required for Shoot Apical Meristem Formation: Class III HD-Zip Genes

Another class of transcription factors, encoded by CLASS III HD-ZIP class genes, has recently been shown to play essential roles in establishing the Shoot Apical Meristem (SAM). Mutants which lack activity of the previously discussed *REV* gene, as well as the related *PHABULOSA* and *PHAVOLUTA* genes, are unable to form an SAM. These genes also promote establishment of adaxial tissues in the leaves. As meristems often form in close association with these tissues (e.g., in the axils of leaves), models have been proposed in which the expression of CLASS III HD-ZIP genes promote what could be termed "central" cell identities. In the stem, the "central" identity would include regions that contain apical initials, as opposed to the organogenic regions of the flanking peripheral zone. In the context of leaf development, this "central identity" would encompass the adaxial tissues, including the asymmetrically positioned xylem elements of the leaf as well as closely associated domains from which axillary meristems arise. This type of model receives further support from studies in which the overexpression of CLASS III HD-ZIP class genes causes SAMs to form in abnormal positions.

The regulation of CLASS III HD-ZIP genes is complex. Their polarized patterns of expression, and the mutant phenotypes that result from perturbing this expression, suggests that the normal patterning of the apical domain depends on radial patterning processes. The proteins encoded by these genes contain START domains, suggesting their activity is regulated by as yet undefined sterol type signalling molecules. The region that encodes the START domain is further distinguished by characteristics that suggest these genes are targets of microRNA (miRNA) mediated repression of transcript levels. An early clue to this mode of regulation was the highly conserved nature of the START domain region, which was apparent not just at the protein level, as might be expected of a ligand-binding region, but also at the nucleotide level, which suggested conserved interactions between nucleic acids. This model has received additional support by the characterization of genes that encode miRNAs that are complementary to the conserved nucleotide sequence motif. The validity of the model is supported further by the effects of "silent" mutations, which alter the nucleic acid sequence of the motif without affecting its coding potential. Mutations of this type would prevent miRNA binding, and thus block the normal down-regulation transcripts in peripheral domains and lead to the observed dominant mutant phenotypes.

The regulation of transcript levels by miRNAs is not completely understood. In some cases, miRNAs mediate the formation of the so-called RNA Induced Silencing Complexes (RISC), which suppress the translation of target transcripts. An alternative type of posttranscriptional regulation has been documented in which the annealing of miRNAs to their mRNAs in the nucleus triggers transcript degradation. More recent data suggests that miRNAs may also mediate transcriptional silencing through chromatin-mediated reduction of gene transcription.

By any of these mechanisms (which can collectively be termed RNAi, for RNA inhibition), one would expect that genes involved in RNAi would play an important role in patterning processes. Indeed, this expectation is met by the mutant phenotypes of genes whose products mediate RNAi processes. These include the Arabidopsis *ARGONAUTE* and *ZWILLE* (*ZLL*) (also known as *PINHEAD*) genes, both of which affect tissue polarity and meristem formation. Although it is unclear precisely how *ZLL* functions in the SAM, mutations that disrupt it function to block the accumulation of STM protein, but have no immediate effect on *STM* transcript levels. Further support for the idea that RNAi-dependent mechanisms are involved in normal SAM function is offered by the recent description of the genes required for normal shoot formation in rice, several of which have been shown to encode several distinct components of RNAi machinery.

Azolla *Root Development*

Our undersanding of the mechanics of meristem function has been enhanced by investigations of the apical meristems of ferns and other primitive vascular plants. The root and shoot apical meristems of these organisms have a single large apical cell, which resides at the centre of the meristem. The apical cell is the ultimate source of all the cells in the meristem, as well as the root or shoot body. Brian Gunning at the Australian National University studied in detail the patterns of cell divisions in roots of the water fern *Azolla*.

Azolla roots are unusual in that they do not grow throughout the life of the plant. The *Azolla* root is a determinate organ, and the root meristem has a genetically determined limit to its growth. This condition is in contrast to most roots, in which the activity of the meristem is indeterminate—growth persisting as long as environmental conditions permit it.

The *Azolla* apical cell, which is shaped something like a pyramid with four more or less triangular sides, divides only 50 to 55 times, cutting off derivative cells in each of its four planes, before it stops dividing entirely. With the exception of the root cap, all the cells in the *Azolla* root can be traced to the apical cell, whose division planes are determined with great precision.

The cells that will become the root body are generated by divisions parallel to the three internal triangular faces of the apical cell, forming a wedge-shaped cell known as a merophyte. Each of the three derivative cells subsequently divides a set number of times to form a sector that makes up one-third of the root axis. The planes and numbers of divisions of the apical cell and each of its derivatives are very precise. We can recognize two different kinds of divisions—formative and proliferative—which differ in the fate of the daughter cells produced. Both daughter cells resulting from a proliferative division have the same fate.

Formative divisions may be either periclinal or anticlinal, but they differ from proliferative divisions in that the two daughter cells have different fates. That is, one of the daughters may remain undifferentiated while the other daughter cell will differentiate as a particular cell type. Formative divisions establish the files of cells that make up the various tissues of the root; proliferative divisions increase the number of cells in each file.

In the root, most formative divisions are periclinal (longitudinal), while proliferative divisions tend to be anticlinal (transverse). Longitudinal, formative divisions within the merophytes give rise to progenitor cells for the files of cells in the root. For example, the division that generates the endodermis and inner cortical cells occurs in the fourth merophyte on both sides of the root, as do the divisions that delineate the vessel element, pericycle, outer cortex, and root hair–epidermis cell files. Once formed, the progenitor cells usually complete several transverse, proliferative divisions before the progeny differentiate.

Shoot Branching

The arrangement of branches constitutes one of the most conspicuous aspects of shoot architecture. During vegetative growth, the formation of branches is guided by a complex interplay between genetically determined programs of development and the environment. As the plant develops, environmental cues are in some way perceived and interpreted to enable adaptive growth responses, ranging from complex shade avoidance responses, to replacement of shoots lost through herbivory. Root systems show a comparable, but usually less appreciated complexity, which, like shoots, reflects interactions between the environment2 s intrinsic programs. Although these programs have the potential to yield an infinite variety of forms, certain common themes suggest basic control mechanisms may be widely conserved among plants. This section examines the phenomenon of branching from several perspectives, beginning with a discussion of its relationship to shoot development and then shifting to a consideration of roots.

In the seed plants, branching is often constrained by phyllotaxy, since the predominant form of branching in most species is axillary (i.e., meristems form at the base of leaves). Superimposed on this basic pattern are further controls that regulate when and where such axillary meristems become active. In many species, axillary buds may lie dormant for long periods, becoming active only when the plant senses favourable growth conditions. Alternatively, buds can be activated by intrinsic factors, such as those associated with the onset of reproductive development.

Formation of the axillary meristem can be linked to the activity of several classes of transcription factors. Seed plants provide examples where secondary shoot systems can form in a variety positions on the plant, ranging from the small plantlets on leaf margins of Kalenchoe, to root-derived tuberous shoots. However, in most plants, such axillary shoots typically form in close association with the adaxial side of the leaf base. Because of this, the pattern of branching is often linked to some degree with the pattern of phyllotaxy.

The earliest stages of axillary meristem development begin with the appearance of a bulge at the leaf primodia, soon after its initiation. *Knotted*-like genes are expressed from this early stage, in contrast to their down-regulation in adjacent leaf primordia. As the axillary develops further, it establishes continuity with the vascular network of the parent growth axis and begins to initiate lateral organs such as leaves or bracts. In many plants, this formative period is followed by a period of dormancy, which can be broken by either intrinsic or external cues.

Although the programs that underlie the formation of axillary buds are undoubtedly complex, some insight into their nature has been offered by the analysis of mutants in which the process is somehow altered. The mutations that affect the polarity of the leaf also influence the pattern of meristem development, suggesting that factors associated with adaxial patterning of the leaf somehow promote meristem formation. Mutations that activate such programs, such as those associated with the dominant alleles of *Phabulosa* and *Phavoluta*, lead to the ectopic formation of axillary meristems in both adaxial and abaxial positions. The related transcription factor, *Revoluta*, which functions in the normal establishment of a polarized vasculature, appears to also play a role. Loss of functions mutations that lead to vascular polarity defects can also block the formation of axillary meristems. The presence of *Start* domains in these Class III HD-ZIP class transcription factors suggest that their activity may be sensitive to an unidentified sterol-like signalling molecule.

Expression analyses suggest that other genes, such as *CUC*, that are required for the formation of the primary SAM in the embryo, are likely to play a similar role for axillary meristems. However, certain types of mutants in which only axillary meristems are affected point towards a novel class of more specific genes. One of these, *lateral suppressor*, was first described in tomato as a mutant in which lateral branches are completely blocked. Cloning of this gene revealed a VHIID-type transcription factor, suggesting that axillary branching is normally dependent on regulation of transcription by the LAS protein. The mechanism that LAS functions in appears to be well-conserved, as LAS-like genes have been described in both Arabidopsis and rice, which have similar branch-deficient mutant phenotypes. Some insights into the potential targets of LAS have been provided by analyses of mutants in Arabidopsis where loss of gene function is linked to a more diffuse pattern of STM expression. Thus, LAS may enable axillary meristem formation by activating genes normally involved in SAM formation. It remains unclear what signals are responsible for the localized expression of LAS.

Once formed, axillary meristems may lie dormant or become activated through a combination of environmental and intrinsic factors, and it appears that many of these responses are determined in part via hormone-signalling pathways. The plant growth hormone ABA, know for its role in promoting seed dormancy, has analogous activities in maintaining the dormancy of axillary buds. Factors that regulate synthesis, turnover, perception, and response of ABA are likely to play important roles in both processes.

Several lines of evidence implicate cytokinins as promoting axillary meristem growth. In classic regeneration experiments, a relatively high ratio of cytokinin to auxin was shown to be essential for regeneration of shoots from undifferentiated callus, whereas low ratios led to the formation of roots. Comparable effects are observed in *Agrobacterium*-infected plants. Plants infected with cytokinin-producing *Agrobacterium tumefaciens* strains develop ectopic shoots, whereas auxin-producing *Agrobacterium rhizogenes* develop ectopic roots. In addition to its role in initiating shoot meristems, cytokinin also appears capable of activating shoots when applied exogenously. Combining cytokinins effects on both the initiation and activation of axillary meristems has the potential to cause massive increases in branching, as suggested by the highly branched phenotype of *supershoot,* which have been linked to increased levels of cytokinin.

The activity of axillary meristems can also be repressed by factors associated with the growth of the primary shoot, a phenomenon known as apical dominance. This type of response can help ensure the survival of the plant where growing shoot tips have been damaged by herbivory or other environmental stresses. Apical dominance phenomena are also likely to be at least partly responsible for some of the complex branching patterns associated with flowering, including the elaboration of characteristic inflorescence architectures, or the release of more proximal vegetative buds to enable perennial growth.

Although the phenomenon of apical dominance is well known and has been exploited throughout the history of agriculture to yield more productive plant forms, its basis still remains incompletely understood. Apical dominance can be inferred when the decapitation of a plant leads to the activation and growth of axillary buds.

As this activation can be suppressed by the application of auxin to the decapitated plant, and since growing shoots are a rich source of this hormone, it is reasonable to suppose that auxin plays some role in apical dominance. Indeed, physiological studies show that auxin is normally transported basipetally from the growing shoot of intact plants via auxin-efflux transporters. An association with auxin is reinforced by the observation that mutants such as *axr1,* which render the plant less-sensitive to auxin, lead to increased branching. However, auxin is normally not efficiently transported into buds, suggesting that its suppression of axillary bud-growth is likely to be mediated indirectly.

The possibility that cytokinin might have a role in mediating apical dominance is suggested by its pattern of synthesis in plants and its growth-promoting activities. Cytokinin is known to be synthesized in roots; from there it is transported acropetally in xylem sap, as well as in nodal regions associated with axillary buds. Furthermore, the inhibitory effects of auxin on bud-growth can be bypassed by exogenous cytokinin, suggesting cytokinin acts downstream of auxin.

The involvement of a novel hormone-signalling pathway in the regulation of bud-growth has been inferred by genetic analyses of branching in Arabidopsis, pea and petunia. In Arabidopsis, a suite of mutants with more axillary branches encoded by the *MAX 1, 2, 3,* and *4* genes has been analysed by a combination of surgical and molecular techniques. Grafting experiments suggest that *MAX 1, 3,* and *4* can act nonautonomously.

For example, the excessive branching of a shoot, which is mutant for any one of these genes, can be rescued by grafting it onto a normal root stock. By contrast, *max2* mutant shoots remain highly branched when grafted. These studies suggested that *MAX1, 3* and *4* are involved in the synthesis of a mobile signal, whereas the *MAX2* gene product mediates perception and response to the signal (e.g., by acting as a receptor). When various mutant versions of *MAX* genes are combined, there is no significant increase in branching, suggesting that these genes act in a linear response pathway rather than acting separately on independent pathways.

Figure: *Phenotypic comparison between (a) wild-type dwarf pea and (b) rms1, and between (c) max4-1 mutants and (d) wild-type Arabidopsis. max4 and rms1 exhibit increased branching, which occurs due to increased outgrowth of lateral buds. Pots are 10 cm in diameter. (Figure kindly supplied by Karim Sorefan.)*

Molecular cloning of the *MAX* genes have born out models suggested by grafting analyses. Both *MAX3* and *MAX 4* encode related gene products that are predicted to act as dioxygenases on carotenoid-type molecules. *MAX1*, by contrast, encodes a cytochrome P450-type enzyme, which is proposed to act downstream of *MAX3* and *MAX 4* to convert the carotenoid product of these genes into a hormone with branch-inhibiting activity (Booker et al. 2005). The identification of this unknown hormone through a combination of biochemical and genetic methods should offer valuable insights into the regulation of bud-growth and may offer new tools for crop improvement.

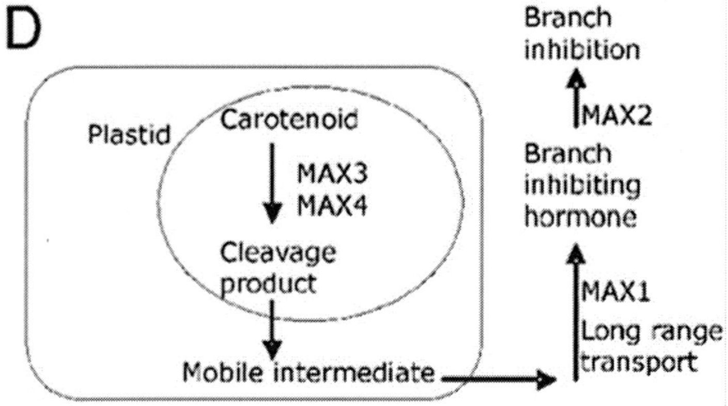

Figure: *Model for proposed functional relationships between* MAX *gene activities in relationship to branching.*

In contrast to *MAX 1, 3,* and *4,* which encode proteins involved in the synthesis of the branch-inhibiting hormone, *MAX2* has been shown to encode an F-box protein. As described in more detail elsewhere, F-box proteins can, when activated, target specific proteins for degradation. In the case of *MAX2,* one can hypothesize one or more proteins that activate bud-growth are normally targeted for destruction by the activated *MAX2* gene product. By this model, the perception of the carotenoid branch-inhibiting hormone would, in some way, activate the *MAX2* gene product. By analogy with the auxin receptors, which have also been shown to encode F-box proteins, it is conceivable that the MAX2 protein itself acts as a receptor, with binding of the branch-inhibiting hormone activating the ability of MAX2 to target specific proteins for degradation. Presumably, these proteins, which would be stabilized in the absence of MAX2 activity, would in some way promote branching.

Chapter

9

Translocation in the Phloem

In classic experiments on the translocation of organic solutes performed by the Italian anatomist Marcello Malpighi in 1686, the bark of a tree was removed in a ring around the trunk. In 1928, T. G. Mason and E. J. Maskell observed that this treatment, called girdling, has no immediate effect on transpiration, since water moves in the xylem, interior to the bark. However, sugar transport in the trunk is blocked at the site where the bark has been removed. Sugars accumulate above the girdle—that is, on the side toward the leaves—and are depleted below the treated region.

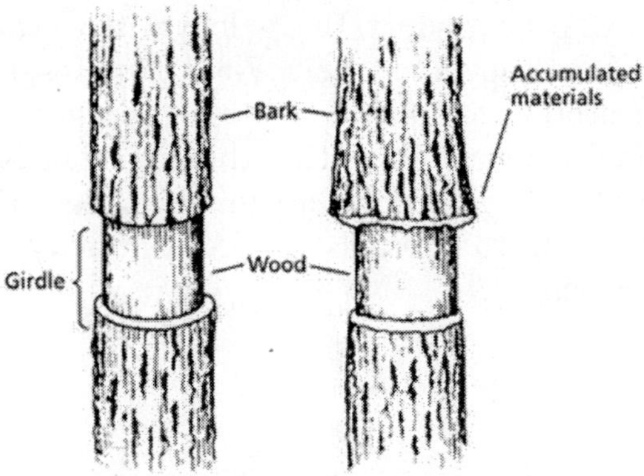

Figure: *Tree trunk immediately after girdling (left) and later (right). Girdling is the removal of the bark of a tree in a ring around the trunk. At right, materials translocated from the leaves have accumulated in the region above the girdle and caused it to swell.*

Eventually the bark below the girdle dies, while the bark above swells and remains healthy. Mason and Maskell concluded that sugar is transported in the bark of the tree and that the sieve elements are the cellular channels of sugar transport. The latter conclusion was based on an observed high correlation between leaf and bark sucrose contents and on the high sucrose concentration calculated to be present in the sieve elements.

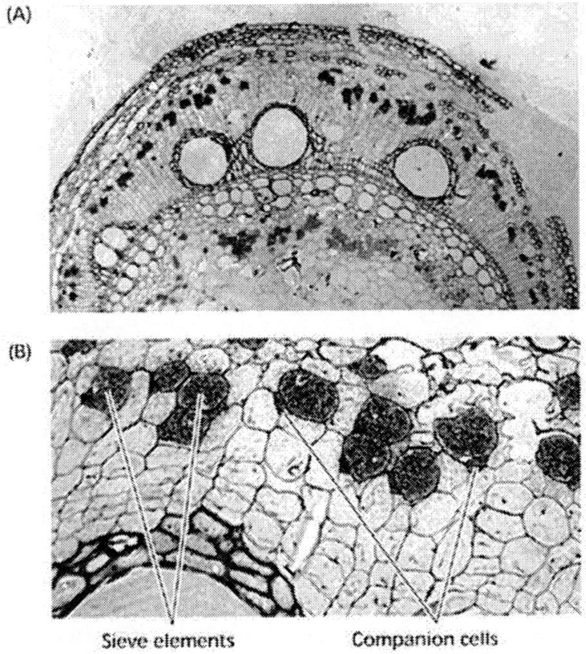

Figure: *Photosynthate from the leaves appears in the sieve elements of phloem in the stem.* $^{14}CO_2$ *was supplied to a source leaf of morning glory (Ipomea nil).* ^{14}C *was incorporated into sugars synthesized in the photosynthetic process, which were then transported to other parts of the plant. The location of the label is revealed in the tissue cross sections by the presence of dark grains on the film. (A) shows a low magnification of the cross section of the stem (50×), revealing dark spots resulting from the silver grains in the film, shown in higher magnification (325×) in (B). The label is confined almost entirely to the sieve elements of the phloem.*

More sophisticated experiments on phloem translocation became possible in the 1940s, when radioactive isotopes became available for scientific research. Labelled organic compounds can be introduced into the plant in various ways. For example, carbon dioxide can be labelled with ^{14}C or ^{11}C and supplied to an intact mature leaf enclosed in a sealed chamber. These two isotopes of carbon, ^{11}C and ^{14}C, have very different half-lives (the half-life is the time required for the radioactivity to decrease by one-half). The half-life for ^{11}C is 20 minutes; that for ^{14}C is 5,600 years.

For this reason, the isotopes are used in different types of experiments. Carbon-14 is useful for long term experiments analysing the compounds in which the radioactivity is present. Carbon-11 would disappear before such an analysis could be made. Carbon-11 is most useful when the researcher wishes to repeat measurements on the same plant. Since only a few hours are needed for the isotope in the plant to disappear, the radioactivity used in one experiment does not interfere with next.

The labelled carbon dioxide is incorporated into organic compounds via the CO_2-fixing reactions of photosynthesis and eventually into transport sugars. In another type of experiment, researchers bypass the photosynthetic reactions by supplying labelled transport sugar, usually sucrose, directly to the leaf.

To identify the pathway of sugar translocation, we must determine the tissue or cellular location of the isotope, usually by means of a technique called autoradiography. In tissue autoradiography, tissue of labelled plants is rapidly frozen, freeze-dried, embedded in paraffin or resin, and sliced into thin sections. The sections are then coated with a film of photographic emulsion.

During a period of exposure, radiation from the radioactive compound fed to the plant exposes the film, and when the film is developed, silver grains appear wherever the label was present in the tissue. A comparison of the tissue section and the pattern of silver grains shows the location of the label in the tissue. In terms of the transport pathway for sugars, the label initially appears in the sieve elements of the phloem, confirming the results of the earlier experiments.

Nitrogen Transport in Soybean

One of the most widely studied species in terms of nitrogen transport is the economically important soybean, a legume that has nitrogen-fixing root nodules. Research has shown that 12 to 48% of the nitrogen needs of the developing seeds of this species are supplied from compounds newly synthesized in the leaves, and 35 to 52% from compounds synthesized in the roots and transferred from xylem to phloem. The remaining nitrogen probably comes from the breakdown of specialized storage proteins in the soybean leaf.

In a recent study, non-nodulated soybean (*Glycine max* (L.) Merr.) plants were cultivated hydroponically under 5 m*M* $NaNO_3$ (ample N supply)) or 0.5 m*M* $NaNO_3$ (deficient N conditions). $^{13}N^-$ or $^{15}N^-$ labelled nitrate was fed to the cut end of the stems, and the accumulation of nitrate-derived N in the pods, nodes and stems was compared. Real-time images of ^{13}N distribution in stems, petioles and pods were obtained using a Positron Emitting Tracer Imaging System for a period of 40 min. The results indicated that the radioactivity in the pods of N-deficient plants was about 10 times higher than that of plants that had sufficient N. The fact that the N translocation into the pods was much faster in N-deficient plants may be due to the strong sink activity of the pods in N-deficient plants.

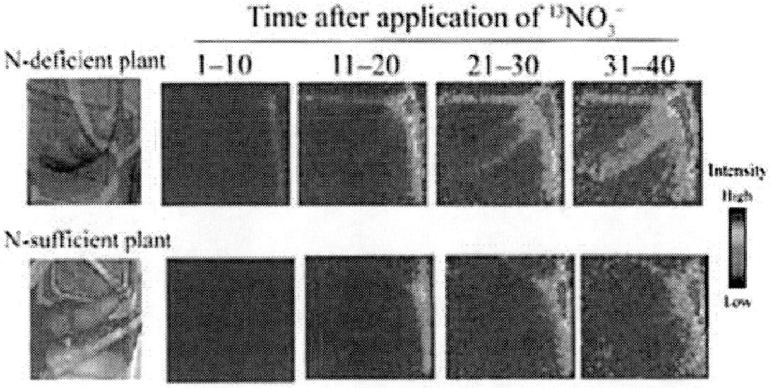

Figure: *Images detected by Positron Emitting Tracer Imaging System around the pod position of the N-deficient and N-sufficient plants. Each image was the average intensity during 10 min. Note that in the 21-30 and 31-40 min. images of the N-deficient pod, the pod image becomes noticeable as radioactivity accumulates in the pod. No comparable image can be seen in the plant with ample N supply.*

Sampling Phloem Sap

Since phloem tissue, like xylem tissue, consists of a mixture of conducting and nonconducting cells, identifying the material translocated in the sieve elements is challenging. Methods based on labelling with radioactive isotopes and direct extraction of the tissue do not distinguish between compounds in transit and those outside the translocation pathway.

The tendency of some species to exude phloem sap from wounds that sever sieve elements has been exploited to obtain relatively pure samples of the translocated material. The driving force for exudation is the high positive pressure in the sieve elements.

However, most species do not exude detectable amounts of phloem sap, because of the sealing mechanisms closing wounds in the phloem. However, exuation rates can be increased by treating the cut surface with the chelating agent EDTA (ethylenediaminetetraacetic acid). Callose synthase—the enzyme that synthesizes the sealing compound, callose—requires the presence of calcium, and since chelating agents bind divalent cations such as calcium very effectively, they lower the calcium concentration available for callose synthase.

The major disadvantage of the wound exudation method is that the fluid collected may not represent the true composition of the translocated material. Contaminants can originate from damaged parenchyma cells or even from the sieve elements themselves, and the sap is generally diluted by water influx from the xylem. Furthermore, the abrupt lowering of turgor pressure in the sieve elements, causes a decrease in the water potential of these cells. As a result, water from the surroundings enters the sieve elements along a water potential gradient, causing the phloem sap to become diluted.

Figure: *Using aphids to tap a single sieve element and collect phloem sap. (A) An aphid (*Longistigma caryae*) feeding on a branch of linden (*Tilia americana*). The high turgor pressure in the sieve element forces the cell contents through the food canal of the aphid and into its gut. The honeydew excreted by the aphid consists of sieve element sap from which selected solutes, especially amino acids, have been removed in the gut of the insect. The high turgor pressure in the sieve element forces the cell contents through the food canal of the aphid and into its gut. The insect in the photo is releasing a droplet of honeydew, which it does about once every 30 minutes. (B) Transverse section through the bark of linden, showing the tips of aphid stylets that had been exuding before they were sectioned. The tips of the stylets are inside a single sieve element; the sheath of saliva secreted by the aphid ends outside the cell.*

The ideal way to collect phloem sap would be to tap into a single sieve element by using a tiny syringe. Fortunately, nature has provided just such a probe: the aphid stylet. Aphids are small insects that feed by inserting their mouthparts, consisting of four tubular stylets, into a single sieve element of a leaf or stem. The high turgor pressure in the sieve element forces the cell contents through the insect's food canal and into its gut, where amino acids are selectively removed and some sugars are metabolized.

The excess sap, still rich in carbohydrates, is then excreted as "honeydew", which can be collected for chemical analysis. Sap can be collected from aphid stylets cut from the body of the insect, usually with a laser, after the aphid has been anesthetized with CO_2. Exudate from severed stylets provides a more accurate picture of the substances present in the sieve elements than does honeydew (aphid excreta), the composition of which has been altered by the insect.

Experimental techniques using aphids have substantial advantages in the study of phloem physiology. Of particular importance is the fact that aphids tap a single sieve element, so there is no problem of contamination from other cell types. Exudation from severed stylets can continue for hours, suggesting that the aphid prevents normal sealing mechanisms from operating or that the drop in pressure that occurs when the sieve element is penetrated is not sufficient to trigger sealing mechanisms.

The disadvantages of using aphids include difficulties in placing the insects at a desired location and severing the stylets without disrupting them. In addition, aphids may induce reactions in the host plant by secreting saliva into the plant tissues. However, the magnitude of these reactions over the period of time necessary to collect sap for analysis is probably negligible.

Monitoring Traffic on the Sugar Freeway

The speed of sugar molecules along the sieve elements of the phloem is known as the mass transfer rate. Measurements of mass transfer rates were among the earliest quantitative determinations made in phloem physiology. The early studies measured either gain in weight by developing fruits or storage organs (sinks) or loss of weight by mature leaves (sources) over a known time interval; the sink experiments are generally the more accurate.

For example, in 1926 Mason and Lewin determined weights of individual tubers of the greater yam (*Dioscorea alata*) over a growing season. The increase in dry weight was divided by the time interval to obtain an average mass transfer rate.

This rate, divided by the average sieve element area per stem, yields a specific mass transfer rate of 5.7 g h^{-1} cm^{-2} of sieve elements. Several similar studies obtained mass transfer rates of 0.2 to 4.8 g h^{-1} cm^{-2} of phloem. If sieve elements are assumed to occupy approximately one-third of the phloem cross-sectional area (a reasonable estimate), these results yield specific mass transfer rates of 0.6 to 14.5 g h^{-1} cm^{-2} for sieve elements.

A more recent technique for determining mass transfer rates by use of radioactive tracers gives rates in excellent agreement with the earlier measurements. In this method the plant is usually pruned to a single mature source leaf and a single immature sink leaf. Carbon dioxide labelled with ^{14}C is supplied to the source leaf in such a way that the rate of supply is equal to the rate of fixation.

The rate of arrival of label in the sink tissue is monitored for the duration of the experiment. The supplied radioactive CO_2 contains a predetermined amount of radioactivity per unit weight of carbon (specific activity). After a certain period of time, the sugars in transit from the source to the sink reach the same specific activity as the supplied CO_2.

Evidence for Apoplastic Loading of Sieve Elements

Early research on phloem loading focused on the apoplastic pathway. Many studies devoted to testing these predictions have provided solid evidence for apoplastic loading in several species. Some of the more important evidence is:

1. Transport sugars are found in the apoplast. Sucrose is the predominant apoplastic sugar in species that transport mainly sucrose in the phloem, for example, sugar beet (*Beta vulgaris*) and broad bean (*Vicia faba*). Treatments and events that alter the rate of translocation from the source leaf also change the flux of sucrose through the apoplast or the apoplastic sucrose level.
2. Sucrose supplied exogenously to a source leaf of sugar beet accumulates in the sieve elements and companion cells of the minor veins, as does sucrose derived from photosynthetic CO_2 fixation. Similar observations have been made in broad bean, pea (*Pisum sativum*), castor bean (*Ricinus communis*), and other species.

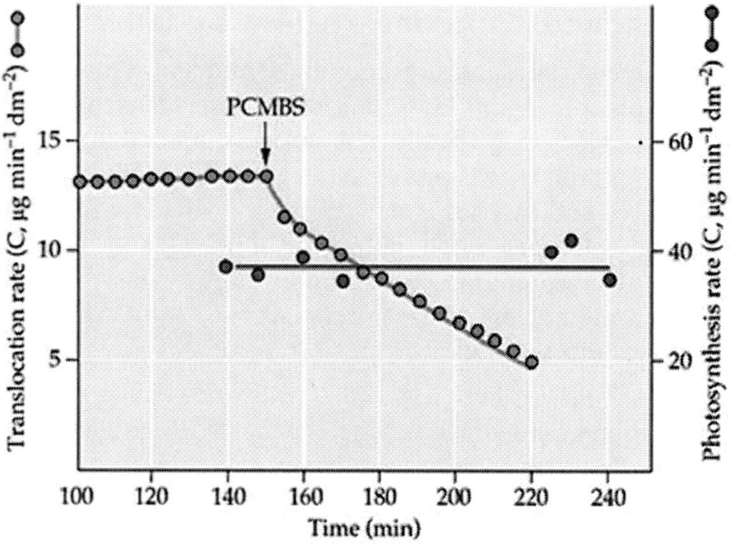

***Figure:** Effects of PCMBS on phloem transport from a sugar beet leaf. Export pf labelled photosynthate from a sugar beet leaf decreases upon addition of PCMBS. Photosynthesis was not affected, indicating that PCMBS did not alter the photosynthetic properties of the leaf.*

3. PCMBS (*p*-chloromercuribenzenesulfonic acid) is a reagent that inhibits the transport of sucrose across plasma membranes but does not enter the symplast. PCMBS inhibits the uptake of sucrose from the apoplast when the sugar is supplied exogenously to sugar beet. Of greater importance, PCMBS also inhibits the export of sucrose synthesized from CO_2 in the mesophyll, implying that short-distance transport in sugar beet normally includes an apoplastic step. Assimilate loading is *also* inhibited by PCMBS in broad bean, pea, and other species.

4. Experiments with transgenic plants further support the interpretation that sucrose moves through the apoplast. (Transgenic organisms carry cloned genes integrated into their DNA by a variety of recombinant-DNA techniques.) Transgenic tomato plants that have a sucrose-cleaving enzyme (invertase) in their apoplast show a very slow growth rate and fail to mobilize the starch in their source leaves during a prolonged dark period. Apparently, hydrolyzing sucrose in the apoplast inhibits phloem loading in source leaves of this species. This result is expected only if the phloem-loading pathway includes an apoplastic step.

Some Substances Enter the Phloem by Diffusion

Many substances, such as organic acids and plant hormones, are found in the phloem sap at lower concentrations than carbohydrates. These substances are probably not actively loaded into the sieve element–companion cell complex but enter the sieve elements via other pathways and mechanisms. They may be taken up directly by diffusion across the phospholipid bilayer of the plasma membrane of the sieve element–companion cell complex or by a passive transporter in the plasma membrane of those cells, or they may diffuse into the sieve elements via the symplast.

Once in the sieve elements, these substances are swept along in the translocation stream by bulk flow, the motive force being generated by the active loading of only certain sugars or amino acids. Many substances not normally found in plants, such as herbicides and fungicides, can be transported in the phloem because of their ability to diffuse through membranes at an intermediate rate.

In other words, they diffuse through membranes rapidly enough to allow considerable accumulation in the sieve elements, but slowly enough that they are not lost from the sieve elements completely before reaching a sink tissue. In addition, compounds that are weakly acidic tend to be "trapped" within the sieve elements, becoming negatively charged in the basic environment (low H^+ concentration) of the sieve elements and thus less likely to diffuse out of the cells across the hydrophobic membrane. Substances that are not transported in the phloem (such as calcium) may not enter the sieve elements at all.

A recent study has characterized the *Arabidopsis* gene *AUX1* (auxin influx carrier) that encodes an auxin permease that appears to functin as an auxin carrier. *AUX1* has been localized in protophloem cells in the root. Gene expression studies have shown that *AUX1* facilitates IAA loading into the leaf vascular transport system, and IAA unloading in the primary root apex and developing lateral root primordium.

Localization of the Sucrose–H^+ Symporter in the Phloem of Apoplastic Loaders

SUC2 is one of the several sucrose transporters that have been localized in the phloem. As shown in Figure below, SUC2 is found in the companion cells.

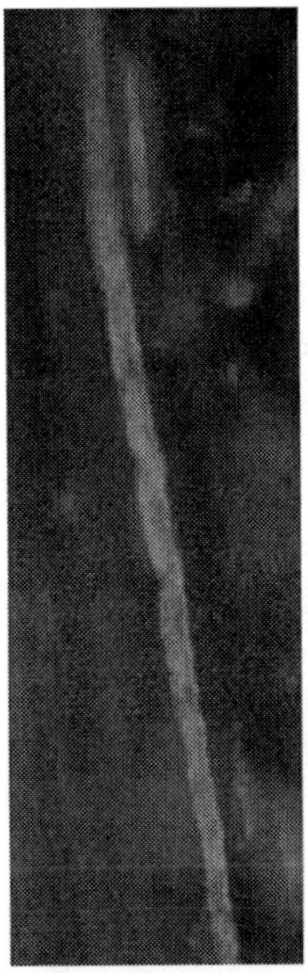

Figure: *This micrograph shows a single companion cell from broad-leaved plantain (Plantago major) stained with two fluorescent dyes. One of the dyes (green) is (indirectly) linked to an antibody that is specific for the PmSUC2 sucrose–H^+ symporter. The second dye (blue) binds to DNA. Since the two dyes are found on a single phloem cell, which is always adjacent to a sieve element, the sucrose symporter is concluded to be located in the companion cell membrane in this species. (From Stadler et al. 1995, courtesy of N. Sauer.) Interestingly, expression of the SUC2 transporter begins in the tip and proceeds to the base in developing leaves during a sink-to-source transition, the same pattern shown by photosynthate export capacity.*

Mutant *Arabidopsis* plants containing transferred DNA insertions in the gene encoding SUC2 have been recently identified by reverse genetics. In the homozygous state, these mutations resulted in stunted growth, retarded development, and sterility. The source leaves of mutant plants contained a great excess of starch, and radiolabeled sugar failed to be transported efficiently to roots and inflorescences. Sucrose transporters are postulated to load photosynthetically produced sucrose into the sieve elements.

Sugars in the Phloem

In symplastic phloem unloading, transport sugars such as sucrose move through the plasmodesmata to the sink cells. In the sink cells, sucrose can be metabolized in the cytosol or the vacuole before being stored or entering metabolic pathways associated with growth of the tissue. When phloem unloading is apoplastic, however, there is an additional opportunity for metabolic change. The transport sugar can be partly metabolized in the apoplast, or it can cross the apoplast unchanged. For example, sucrose can be hydrolyzed into glucose and fructose in the apoplast by invertase, and glucose and/or fructose would then enter the sink cells. The fact that many monosaccharide transporters have been localized mainly in sink tissues supports the possible existence of this pathway.

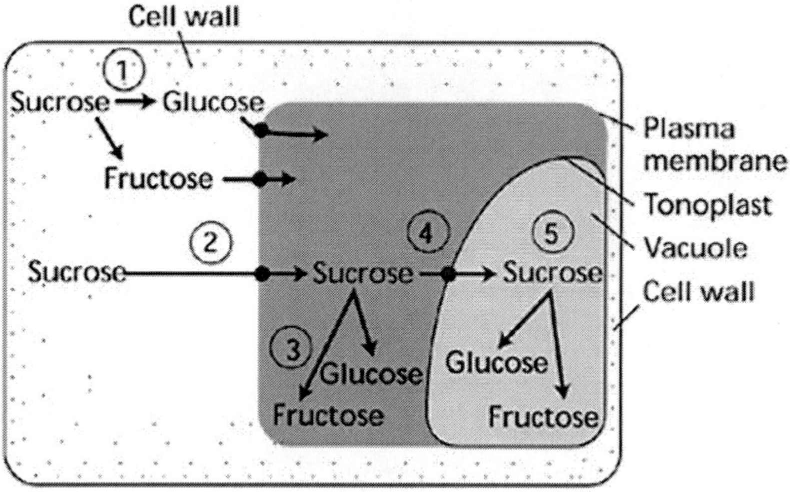

Figure: *The possible fates of sucrose unloaded apoplastically in sink tissues. (1) Sucrose that enters the apoplast can be split into glucose and fructose by a wall invertase before entering a cell from a sink tissue, or (2) sucrose can be taken up into the cell unaltered. (3) Once in the symplast of the cell from the sink tissue, sucrose can be split into glucose and fructose by a cytoplasmic invertase, or (4) sucrose can enter the vacuole unaltered. (5) Once in the vacuole, sucrose can be split into glucose and fructose by a vacuolar invertase, or it can remain unaltered.*

A recent study with potato plants has shown that apoplastic unloading predominated in elongating stolons. Stolons are underground lateral shoots that grow from the main stem of the potato plant, characterized by elongated internodes and hooked apical tips, that form tubers at their apices in response to environmental signals. When tuberization started in stolons, phloem unloading shifted from apoplastic to symplastic transport. Histochemical analysis of potato lines transformed with the promoter of an apoplastic invertase gene (*invGE*) linked to a reporter gene showed invertase activity in the elongating stolon, associated with apoplastic unloading. In the developing tuber, apoplastic loading and invertase activity was observed in a small apical region, which was the apical area of the stolon progressively engulfed by the swelling subapical regions during tuberization. Most of the tuber showed symplastic unloading and lacked expression of the invertase gene.

Energy Requirements for Unloading in Developing Seeds and Storage Organs

Developing seeds have proven to be a most interesting system in which to study unloading processes. In legumes such as soybean, the embryo can be removed from the seed coat. In this way, unloading from the seed coat into the apoplast can be studied without the influence of the embryo, and uptake into the embryo can also be investigated separately. Studies with legumes have shown that both entry of sucrose into the apoplast and uptake into the embryo are mediated by transporters and are active. In cereals like wheat, only uptake into the embryo is active; the loss of sucrose (sucrose efflux) from the maternal tissues is passive (down the concentration gradient), because the subsequent active step keeps the sucrose concentration in the apoplast low.

In corn, the cell wall invertase helps maintain a low apoplastic sucrose concentration by splitting the disaccharide into monosaccharides. In general, sugar–proton symport mechanisms appear to function in the uptake of sugars from the apoplast, as in sucrose uptake into the soybean embryo. Storage organs often accumulate sugars to high concentrations, for example, in sugar beet taproot and sugarcane stem. This sugar accumulation requires active membrane transport, since energy is required to move sugars into storage compartments against a concentration gradient. Sugar transport into the vacuoles of storage cells such as those of sugar beet is thought to be accomplished by a sucrose–proton antiport. In this case, a vacuolar H^+-ATPase pumps protons into the vacuole; the antiport carrier then moves sucrose into the vacuole in exchange for protons, which exit the vacuole down their electrochemical-potential gradient.

Possible Mechanisms Linking Sink Demand and Photosynthetic Rate in Starch Storers

Photosynthetic rate and sink demand are linked processes, particularly in species which store starch. Increased sink demand is correlated with increased photosynthesis, and vice versa. An accumulation of photosynthate (starch, sucrose, or hexoses), in the source leaf could be one of the signals coupling photosynthetic rate with sink demand. Possible mechanisms include:

1. *Inhibition by starch:* When sink demand is low, high starch levels in the source could physically disrupt the chloroplasts, interfere with CO_2 diffusion, or block light absorption. Little evidence supports this hypothesis.
2. *Phosphate availability:* When sink demand is low, photosynthesis could be restricted by a lack of free orthophosphate in the chloroplast. Under conditions of low demand by the sink, sucrose synthesis is usually reduced, and less phosphate is thus available for exchange with triose phosphate from the chloroplast (via the phosphate translocator). If starch synthesis, which releases orthophosphate in the chloroplast, could not release phosphate fast enough, a deficiency in phosphate would ensue. ATP synthesis and thus CO_2 fixation would decline. A study with potato plants transformed with antisense DNA to the phosphate translocator provides support for this hypothesis. The transformed plants, which displayed reduced phosphate translocator activity, allocated proportionately more carbon into starch and less into sucrose. These effects were accompanied by a reduction in the light-and CO_2-saturated rates of photosynthesis in young plants.
3. *Regulation by sugars:* High sugar levels decrease the transcription rate and expression of genes for many photosynthetic enzymes. The changes in gene expression occur over the same time frame as the source adjustments already described. For example, in source leaves of spinach (*Spinacia oleracea*), mRNA for several photosynthetic enzymes decreased when soluble carbohydrates accumulated as a result of inhibition of export from the leaf. Although transcript levels began to decline almost immediately, changes in photosynthetic enzyme activity were apparent only after several days. In this species at least, photosynthesis appeared to be inhibited because of changes in gene expression, not because of phosphate limitation.

Chapter

10

Overview of Plant Metabolism

Plant Cells

The Plant Kingdom

Since the time of Aristotle (384–322 B.C.E.), biologists have sought to classify organisms. At first the purpose was ease of identification ("artificial" classification schemes). Carolus Linnaeus (1707–1778), arguably the greatest of the pre-modern Naturalists, sought to classify plants and other organisms according to affinity groups that reflected the mind of the Creator. Later, after Darwin, the goal of classification was to show evolutionary relationships ("natural" classification schemes).

For the past 150 years, biologists have emphasized natural systems of classification and have attempted to define morphological criteria that reveal evolutionary relationships.

We now know that morphology, the form and structure of organisms, is the end product of the actions of genes. Virtually all of the information needed to form a complete organism is encoded in its DNA sequences, both nuclear and cytoplasmic (mitochondria and chloroplasts). DNA sequence analysis has thus provided evolutionary biologists with a powerful new tool for arriving at a truly natural classification system.

On the basis of phylogenetic analyses of highly conserved DNA sequences, living organisms have been divided into three major domains: Bacteria, Archaea, and Eucarya.

1. The common ancestor of all the organisms first gave rise to the Bacteria and the common ancestor of the Archaea and the Eucarya.
2. The Archaea branch off from the Eucarya lineage.
3. The Eucarya common ancestor acquires mitochondrial endosymbiont.
4. A heterogeneous group of eukaryotes called protists branch off the lineage leading to plants, fungi, and animals.
5. The common ancestor of fungi and animals form a branch, followed by a divergence into the fungal and animal lineages.
6. The common ancestor of plants, green algae, red algae, and glaucophytes acquires chloroplast endosymbiont (a cyanobacterium).
7. The three lineages of glaucophytes, red algae, and green algae diverge.
8. Various lineages of protists acquire chloroplasts via green or red algal endosymbionts.
9. The earliest branch of green algae diverge.
10. The later branch of green algae diverge.
11. The remaining lineage leads to plants.

The Eucarya include the eukaryotes, organisms whose cells contain a true nucleus. The Bacteria, or eubacteria, which include the cyanobacteria, lack a true nucleus and are therefore prokaryotic.

The Archaea, or archaebacteria, are also prokaryotic, but they differ from the Bacteria: Besides their morphological and biochemical differences, they are often adapted to extreme environments, such as sulfur hot springs or saline ponds. Phylogenetic studies have indicated that the Archaea and Eucarya split after the Bacteria separated from the common ancestor. Thus Archaea and Eucarya represent sister groups. This closer relation between Archaea and Eucarya is reflected in their similar promoter structures and RNA polymerases, the presence of histones, and many other characteristics.

Fungi were formerly classified as algal-like plants that had lost their chloroplasts. However, fungi and animals branched off from the Eucarya lineage *before* the appearance of choloroplasts. They are thus more closely related to animals than to plants. Fungi are heterotrophic; that is, they depend on other organisms for their food, and they satisfy their nutritional needs by absorbing inorganic ions and organic molecules from the external environment. Most fungal species are filamentous and possess cell walls made of chitin, the same substance that is found in insect exoskeletons.

Bryophytes are small (rarely more than 4 cm in height), very simple land plants, and the least abundant in terms of number of species and overall population. Bryophytes do not appear to be in the direct line of evolution leading to the vascular plants; rather, they seem to constitute a separate minor branch.

Bryophytes include mosses, liverworts, and hornworts. These small plants have life cycles that depend on water during the sexual phase. Water facilitates fertilization, the fusion of gametes to produce a diploid zygote, a feature also seen in the algal precursors of these plants.

Bryophytes are like algae in other respects as well: They have neither true roots nor true leaves, they lack a vascular system, and they produce no hard tissues for structural support. The absence of these structures that are important for growth on land greatly restricts the potential size of bryophytes, which, unlike algae, are terrestrial rather than aquatic.

The ferns represent the largest group of spore-bearing vascular plants. In contrast to the bryophytes, ferns have true roots, leaves, and vascular tissues, and they produce hard tissues for support.

These architectural features enable ferns to grow to the size of small trees. Although ferns are better adapted to the drying conditions of terrestrial life than bryophytes are, they still depend on water as a medium for the movement of sperm to the egg. This dependence on water during a critical stage of their life cycle restricts the ecological range of ferns to relatively moist habitats.

The most successful terrestrial plants are the seed plants. Seed plants have been able to adapt to an extraordinary range of habitats. The embryo, protected and nourished inside the seed, is able to survive in a dormant state during unfavourable growing conditions such as drought. Seed dispersal also facilitates the dissemination of the embryos away from the parent plant.

Another important feature of seed plants is their mode of fertilization. Fertilization in seed plants is brought about by wind-or insect-mediated transfer of pollen, the gamete-producing structure of the male, to the sexual structure of the female, the pistil. Pollination is independent of external water, a distinct advantage in terrestrial environments. Many seed plants produce massive amounts of woody tissues, which enable them to grow to extraordinary heights. These features of seed plants have contributed to their success and account for their wide range.

There are two categories of seed plants: gymnosperms (from the Greek for "naked seed") and angiosperms (based on the Greek for "vessel seed," or seeds contained in a vessel). Gymnosperms are the less advanced type; about 700 species of gymnosperms are known. The largest group of gymnosperms is the conifers ("cone-bearers"), which include such commercially important forest trees as pine, fir, spruce, and redwood.

Two types of cones are present: male cones, which produce pollen, and female cones, which bear ovules. The ovules are located on the surfaces of specialized structures called cone scales. After wind-mediated pollination, the sperm reaches the egg via a pollen tube, and the fertilized egg develops into an embryo. Upon maturation, the cone scales, which are appressed during early development, separate from each other, allowing the naked seeds to fall to the ground.

Angiosperms, the more advanced type of seed plant, first became abundant during the Cretaceous period, about 100 million years ago. Today, they dominate the landscape, easily outcompeting their cousins, the gymnosperms. About 250,000 species are known, but many more remain to be characterized. A typical angiosperm life cycle, that of *Zea mays* (corn). The major innovation of the angiosperms is the flower; hence they are referred to as *flowering plants*. There are other anatomical differences between angiosperms and gymnosperms, but none so crucial and far-reaching as the mode of reproduction.

Flower Structure and the Angiosperm Life Cycle

The flower consists of several leaflike structures attached to a specialized region of the stem called the receptacle. Sepals and petals are the most leaflike. Petals have the primary function of attracting insects to serve as pollinators, accounting for their often showy and brightly coloured appearance.

The stamen is the male sexual structure, and the pistil is the female sexual structure. The pistil is composed of one or more united carpels; the pistil, or in some flowers a whorl of pistils, is sometimes referred to as the gynoecium. The stamen consists of a narrow stalk called the filament and a chambered structure called the anther. The anther contains tissue that gives rise to pollen grains. The pistil consists of the stigma (the tip where pollen lands during pollination), the style (an elongated structure), and the ovary. The ovary, the hollow basal portion of the pistil, completely encloses one or more ovules. Each ovule, in turn, contains an embryo sac, the structure that gives rise to the female gamete, the egg.

After landing on the stigma, the pollen grain germinates to form a long pollen tube, which penetrates the tissues of the style and ultimately enters the cavity of the ovary, which houses the ovule. Within the ovary, the pollen tube enters the ovule and deposits two haploid sperm cells in the embryo sac. One sperm cell fuses with the egg to produce the zygote; the other typically fuses with the two polar nuclei to produce a specialized storage tissue termed the endosperm, which provides nutrients to the growing embryo.

Endosperm tissue also provides the bulk of the worldrquotes food supply in the form of cereal grains. As in conifers, in angiosperms the outer tissues of the ovule harden into a protective seed coat. Angiosperm seeds have a second layer of protective tissues, the fruit. The fruit consists of the ovary wall and, in some cases, receptacle tissue.

Angiosperms are divided into two major groups; dicotyledons (dicots) and monocotyledons (monocots). This distinction is based primarily on the number of cotyledons, or seed leaves. In addition, the two groups differ with respect to other anatomical features, such as the arrangement of their vascular tissues, and their floral structure.

As the dominant plant group on Earth, and because of their great economic and agricultural importance, angiosperms have been studied much more intensively than other types of plants. Plant physiologists have focused on a relatively small number of species that represent convenient experimental systems for the study of specific phenomena.

Therefore, while we focus on these famous few, it is important to keep in mind the tremendous diversity of form and function that exists within the angiosperms, and the even greater diversity of form and function that is found within the plant kingdom as a whole.

Plant Tissue Systems: Dermal, Ground, and Vascular

Dermal tissue. The epidermis is the dermal tissue of young plants undergoing primary growth. It is generally composed of specialized, flattened polygonal cells that occur on all plant surfaces. Shoot surfaces are usually coated with a waxy cuticle to prevent water loss and are often covered with hairs, or trichomes, which are epidermal cell extensions.

Pairs of specialized epidermal cells, the guard cells, are found surrounding microscopic pores in all leaves. The guard cells and pores are called stomata (singular stoma), and they permit gas exchange (water loss, CO_2 uptake, and O_2 release or uptake) between the atmosphere and the interior of the leaf.

The root epidermis is adapted for absorption of water and minerals, and its outer wall surface typically does not have a waxy cuticle. Extensions from the root epidermal cells, the root hairs, increase the surface area over which absorption can take place.

Ground tissue. Making up the bulk of the plant are cells termed the ground tissue. There are three types of ground tissue: parenchyma, collenchyma, and sclerenchyma.

- Parenchyma, the most abundant ground tissue, consists of thin-walled, metabolically active cells that carry out a variety of functions in the plant including photosynthesis and storage
- Collenchyma tissue is composed of narrow, elongated cells with thick primary walls. Collenchyma cells provide structural support to the growing plant body, particularly shoots, and their thickened walls are nonlignified so they can stretch as the organ elongates. Collenchyma cells are typically arranged in bundles or layers near the periphery of stems or leaf petioles.
- Sclerenchyma consists of two types of cells, sclereids and fibres Both have thick secondary walls and are frequently dead at maturity. Sclereids occur in a variety of shapes, ranging from roughly spherical to branched, and are widely distributed throughout the plant. In contrast, fibres are narrow, elongated cells that are commonly associated with vascular tissues. The main function of sclerenchyma is to provide mechanical support particularly to parts of the plant that are no longer elongating.

In the stem, the pith and the cortex make up the ground tissue. The pith is located within the cylinder of vascular tissue, where it often exhibits a spongy texture because of the presence of large intercellular air spaces.

If the growth of the pith fails to keep up with that of the surrounding tissues, the pith may degenerate, producing a hollow stem. In general, roots lack piths, although there are exceptions to this rule. In contrast, the cortex, which is located between the epidermis and the vascular cylinder, is present in both stems and roots.

At the boundary between the ground tissue and the vascular tissue in roots, and occasionally in stems, is a specialized layer of cortex known as the endodermis. This single layer of cells originates from cortical tissue at the innermost layer of the root cortex and forms a cylinder that surrounds the central vascular tissue, or stele.

Early in root development, a narrow band composed of the waxy substance suberin is formed in the cell walls circumscribing each endodermal cell. These suberin deposits, called Casparian strips, form a barrier in the endodermal walls to the intercellular movement of water, ions, and other water-soluble solutes to the vascular cells. Leaves have two interior layers of ground tissue that are collectively known as the mesophyll.

The palisade parenchyma consists of closely spaced, columnar cells located beneath the upper epidermis. There is usually one layer of palisade parenchyma in the leaf. Palisade parenchyma cells are rich in chloroplasts and are a primary site of photosynthesis in the leaf. Below the palisade parenchyma are i rregularly shaped, widely spaced spongy mesophyll cells. The spongy mesophyll cells are also photosynthetic, and the large spaces between these cells allow diffusion of carbon dioxide. The spongy mesophyll also contributes to leaf flexibility in the wind, and this flexibility facilitates the movement of gases within the leaf.

Vascular tissues: xylem and phloem. The vascular tissue is composed of two major conducting systems: the xylem and the phloem. The xylem transports water and mineral ions from the root to the rest of the plant.

The phloem distributes the products of photosynthesis and a variety of other solutes throughout the plant. The tracheids and vessel elements are the conducting cells of the xylem.

Both of these cell types have elaborate secondary-wall thickenings and lose their cytoplasm at maturity; that is, they are dead when functional. Tracheids overlap each other, whereas vessel elements have open end walls and are arranged end to end to form a larger unit called a vessel. Other cell types present in the xylem include parenchyma cells, which are important for the storage of energy-rich molecules and phenolic compounds, and sclerenchyma fibres.

The sieve elements and sieve cells are responsible for sugar translocation in the phloem. The former are found in angiosperms; the latter perform the same function in gymnosperms. Like vessel elements, sieve elements are often stacked in vertical rows, forming larger units called sieve tubes, whereas sieve cells form overlapping arrays. Both types of conducting cells are living when functional, but they lack nuclei and central vacuoles and have relatively few cytoplasmic organelles.

Substances are translocated from sieve cell to sieve cell laterally through circular or oval zones containing enlarged pores, called sieve areas. In contrast, sieve tubes translocate substances through large pores in the end walls of the sieve elements, called sieve plates. Sugar movement through sieve tubes is more efficient and rapid than through sieve cells and represents a more evolutionarily advanced mechanism.

Sieve elements are associated with, and depend on, densely cytoplasmic parenchyma cells called companion cells. The analogous cells adjacent to the sieve cells of gymnosperms are called albuminous cells. Companion cells provide proteins and metabolites necessary for the functions of the sieve tube elements. In addition, the phloem frequently contains storage parenchyma and fibres that provide mechanical support.

The Structures of Chloroplast Glycosylglycerides

The structures of the three plant glycosylglycerides, monogalactosyldiacylglycerol (MGDG), digalactosyldiacylglycerol (DGDG) and sulfoquinovosyldiacylglycerol (SQDG). The two galactolipids, MGDG and DGDG, are uncharged but polar; the sulfolipid, SQDG, is negatively charged as well as being polar.

The Multiple Steps in Construction of the Cell Plate Following Mitosis

During Mitosis in plants, the daughter chromosomes separate during anaphase and migrate toward their respective poles. In contrast to animal cells, the spindle poles of plant cells is quite diffuse, so that the chromosomes move in nearly parallel fashion.

At the end of teleophase, cytokinesis begins with the formation of the cell plate. This process involves fusion of many small secretory vesicles, and attachment of the resulting structure to the plasma membrane. A diagram of the steps involved in cell plate formation.

In the first step, Golgi vesicles, some of which are interconnected via *fusion tubes,* aggregate in the spindle midzone area. This structure is called the fusion tube network (FTN). The transition from the first to the second stage of cell plate formation can be inhibited by caffeine. (B) Formation of a tubulo-vesicular network (TVN).

The contents of the vesicles, mainly pectins, represent the precursors from which the new middle lamella is assembled outside the cell. In the next stage, vesicle fusion increases, forming a tubulo-vesicular network (TVN), and the membranes become coated with either clathrin or other proteins. (C) In the third stage, the central region of the growing cell plate forms a tubular network (TN), with vesicle fusion occurring at the growing edges where the remaining microtubules are located. (D) In the final stage, the cell plate contacts and adheres to the plasma membrane of the parent cell.

At the same time the tubular network expands to form a fenestrated sheet. (E) At the end of mitosis, the phragmoplast disappears, the cell enters interphase, and microtubules reappear in the cytosol near the plasma membrane, where they play a role in the deposition of cellulose microfibrils during cell wall growth. (F) Electron micrograph of a cell plate forming in a root tip of a beet, (*Beta vulgaris*) (10,000×) MT, microtubule; VE, vesicles; N, nucleus; NE, nuclear envelope; P, cell plate.

Water and Plant Cells

Calculating Capillary Rise

We learn on page 36 of the textbook that the water properties of cohesion, adhesion and surface tension, give rise to the phenomenon of capillarity, the movement of water for small distances up a capillary tube. The smaller the tube radius, the higher the capillary rise. How far the water will move may be calculated using the following formula:

$$\text{Capillary rise} = \frac{14.9 \times 10^{-6}\,\text{m}^2}{\text{radius}}$$

where both capillary rise and radius are expressed in meters.

For a xylem vessel with 25 μm radius, the capillary rise is about 0.6 m. This distance is much too small to be significant for water transport up tall trees.

Fibrous materials such as cell walls can act like wicks to draw water by capillarity from nearby xylem.

This capillary action ensures that cell wall surfaces that are directly exposed to the air, such as those in leaf mesophyll, remain wetted and do not dry out. Because the cell wall capillaries have a tiny radius, about 10 m^{-8}, very large physical forces can be generated in the water just below the evaporative surfaces of cell walls.

Calculating Half-Times of Diffusion

Extremely Slow over Long Distances

From Fick's law, one can derive an expression for the time it takes for a substance to diffuse a particular distance.

If one defines conditions such that all the solute molecules are concentrated at the starting position, then the concentration front moves away from the starting position, as shown for a later time point. As the substance diffuses away from the starting point, the concentration gradient becomes less steep (Δc_s decreases) and thus net movement becomes slower.

The time it takes for the substance at any given distance from the starting point to reach one-half of the concentration at the starting point ($t_c = ½$) is given by the following equation:

$$t_{c=1/2} = \frac{(\text{dis tan ce})^2}{D_S} K$$

where K is a constant and D_s is the diffusion coefficient. The above equation shows that the time required for a substance to diffuse a given distance increases in proportion to the *square* of that distance. Let us consider two numerical examples. First, how long it would take a small molecule to diffuse across a typical cell? The diffusion coefficient for a small molecule like glucose is about 10^{-9} m^2 s^{-1}, and the cell length may be 50 μm. Thus, for this example:

$$t_{c=1/2} = \frac{(50 + 10^{-6} \text{m})^2}{10^{-9} \text{m}^2 \text{s}^{-1}} = 2.5\text{s}$$

This calculation shows that small molecules diffuse over cellular dimensions rapidly. What about diffusion over longer distances? Calculating the time needed for the same substance to move a distance of 1 m (e.g., the length of a corn leaf), we find:

$$t_{c=1/2} = \frac{(1\text{m})^2}{10^{-9} \text{m}^2 \text{s}^{-1}} = 10^9 \text{s} \approx 32 \text{ year}$$

a value that exceeds by orders of magnitude the life span of a corn plant, which lives only a few months. This shows that diffusion in solutions can be effective within cellular dimensions but is far too slow for mass transport over long distances. Diffusion is of great importance as a driving force for the water vapour lost from leaves, because the diffusion coefficient for a molecule in air is much greater than in aqueous solutions.

Alternative Conventions for Components of Water Potential

Students planning further study of plant water relations should note that the components of water potential defined in the text are sometimes given different names and symbols. In particular, the equation

$$\psi_w = \psi_s + \psi_p$$

(textbook Equation 3.6) is often replaced by the following equivalent equation:

$$\psi_w = \pi + P$$

In this alternative convention, *P* is the same as Ψ_p. It is the hydrostatic pressure of the solution, and may be positive, as in turgid cells, or negative, as in xylem water.

The symbol π is called osmotic pressure and is the negative of Ψ_s. That is, π has positive values, and Ψ_s has negative values. "Osmotic pressure" is the term that physical chemists, zoologists, and many others use to denote the effect of dissolved solutes on the free energy of water. Most handbooks of physics and chemistry use the term "osmotic pressure" and the symbol π.

The negative sign in front of π in the equation above accounts for the reduction in water potential (Ψ_w) by dissolved solutes. Thus $\Psi_s = -\pi$. A very interesting, if somewhat unconventional, account of the history and physical meaning of osmotic pressure is given by Hammel and Scholander (1976).

Unfortunately, some authors have mixed the conventions for Ψ_s and π, leading to unnecessary confusion about what is meant by the symbol π. Thus, π is sometimes incorrectly called osmotic potential instead of osmotic pressure, and it may be used either as a positive quantity or as a negative quantity.

Regarding matric potential, it is usually designated by the symbol *t*.

Can Negative Turgor Pressures Exist in Living Cells?

One of the challenging aspects of understanding plant water relations is the remarkable range of pressures, both positive and negative, that occur within the bodies of plants.

Negative pressures (or tensions), which depend upon the cohesive strength of water coupled with the strength of lignified cell walls to resist deformation, play an important role in water transport through the xylem. Positive pressures, which depend upon the semipermeable nature of the plasma membrane and the elastic nature of primary cell walls, occur in all hydrated living plants cells but can be especially large in sieve tubes and guard cells.

Living plant cells are typically assumed to have only positive pressures. However, there appears to be no reason that negative pressures could not also occur within the cytoplasm of living plant cells. This web topic explores the existence and potential role of negative turgor pressures in plants.

When a cell loses water in air, turgor declines and solute concentrations increase. At the turgor loss point ($\psi_p = 0$), the hydrostatic pressure in the cell sap is equal to atmospheric pressure, meaning that no net force is exerted on the cell wall.

If water continues to be lost from the cell, the pressure within the cytoplasm drops below atmospheric pressure, resulting in a force imbalance that collapses the cell wall. The deformation of living cells upon desiccation is called cytorrhysis. Note that the plasma membrane is pressed against the cell wall throughout desiccation (i.e., plasmolysis does not occur) because the hydrostatic pressure in cytoplasm remains greater than the hydrostatic pressure in apoplast.

In the living cells described above, a decrease in cell water potential below the turgor loss point is balanced by an increase in solute concentration as the volume of the cell decreases.

Thus, true "negative" pressures do not develop. In contrast, xylem conduits have rigid cell walls that resist deformation, allowing them to sustain negative pressures without imploding. This raises the question of what happens when water is lost from living cells that have thick walls or are embedded in a rigid matrix of cells (e.g., xylem parenchyma). Might these cells resist deformation and thus develop negative turgor pressures?

It is important to emphasize that this is a controversial area, due in large part to the absence of direct measurements of negative turgor pressures within cells. However, before reviewing the evidence for negative turgor pressures, it is worthwhile to consider what physiological effects might result from the development of such negative pressures in living cells.

The major outcome of negative turgor is that it allows stiff-walled cells to decrease in water potential without undergoing major changes in cell volume or osmotic concentration. Because cytorrhysis might cause physical damage to the wall and/or cell membranes, while the high concentrations of solutes resulting from the reduction in cell volume might adversely affect the conformation of membranes and proteins, it is possible to imagine a physiological role or benefit for the development of negative turgor pressures in plant cells.

We now turn this around and ask if there are any downsides to the generation of negative turgor pressures in living cells? In the xylem, the primary issue associated with negative pressures is cavitation; could this also be a concern for living cells?

The primary mechanism for cavitation in plants is air seeding, reflecting the fact that the probability of the de novo formation of gas voids in water (either by homogeneous or heterogeneous nucleation) is extremely low. In air seeding, air is sucked in through the cell wall, where it then "nucleates" the transition to the vapour phase.

For air seeding to occur in a living cell experiencing negative pressures, air would have to be pulled through the very small pores of the cell wall. While one can imagine this happening, the movement of air across the cell wall would result in plasmolysis and thus the immediate release of any tension in the cytoplasm because the plasma membrane is not capable of withstanding any significant pressure.

Other costs associated with the ability of living cells to generate negative pressures include the metabolic costs of producing rigid cell walls, as well as any limitations lignified walls might place on physiological function. In addition, a strategy for avoiding cell damage due to desiccation via the generation of negative pressures might impose limitations on cell size. The strength of a cell to withstand collapse (and thus generate negative pressures) is inversely proportional to cell size.

Evidence for negative turgor pressure in plants is limited, in part reflecting the fact that few researchers have devoted much attention to this issue. Living cells with flexible (unlignified) cell walls deform relatively easily upon desiccation and thus do not appear to support negative pressures. However, measurements of the forces needed to collapse cell walls suggest that living cells with thick walls can withstand forces in the range of 1.0 MPa.

Visual examination of tissues adapted to withstand very cold temperatures also provides indirect evidence for negative turgor. For example, frozen ray parenchyma cells often do not exhibit significant deformation, despite the very strong desiccatory effects (low water potential) of extracellular ice. One can hypothesize that the existence of negative pressures within these cells acts to balance the water potential gradient across the cell membrane (i.e., between the cytoplasm and ice formed within the apoplast).

In summary, negative turgor pressure remains an intriguing but little-studied area of plant water relations. While it does not appear to form in cells such as the leaf mesophyll, that can deform easily, its existence in living cells with either lignified cell walls or where the cell is embedded in a matrix of lignified tissues (e.g., living cells within wood) cannot be ruled out. The potential benefits of negative turgor pressures in terms of preventing mechanical and osmotic damage associated with severe desiccation makes.

The Matric Potential

In discussions of soils, seeds, and cell walls, one often finds reference to yet another component of Ψ_w: the matric potential (Ψ_m). The matric potential is used to account for the reduction in free energy of water when it exists as a thin surface layer, one or two molecules thick, adsorbed onto the surface of relatively dry soil particles, cell walls, and other materials.

The matric potential does not represent a new force acting on water, because the effect of surface interactions can theoretically be accounted for by an effect on Ψ_s and Ψ_p. In dry materials, however, this surface interaction effect often cannot easily be separated into Ψ_p and Ψ_s components in dry materials, so they are frequently bulked together and designated as the matric potential.

It is generally not valid to add Ψ_m to independent measurements of Ψ_s and Ψ_p to arrive at a total water potential. This is particularly true for water inside hydrated cells and cell walls, where matric effects are either negligible or they are accounted for by a reduction in Ψ_p. For instance, the negative pressure in water held by cell wall microcapillaries at the evaporative surfaces of leaves, is sometimes described as a wall matric potential. Care is needed to avoid inconsistencies when accounting for this physical effect in definitions of Ψ_p, Ψ_s, and Ψ_m.

Measuring Water Potential

Plant scientists have expended considerable effort in devising accurate and reliable methods for evaluating the water status of a plant. Four instruments that have been used extensively to measure

Ψ_w, Ψ_s, and Ψ_p are described here: psychrometer, pressure chamber, cryoscopic osmometer, and pressure probe.

Psychrometer (Ψ_w Measurement)

Psychrometry (the prefix "psychro-" comes from the Greek word *psychein*, "to cool") is based on the fact that the vapour pressure of water is lowered as its water potential is reduced. Psychrometers measure the water vapour pressure of a solution or plant sample, on the basis of the principle that evaporation of water from a surface cools the surface.

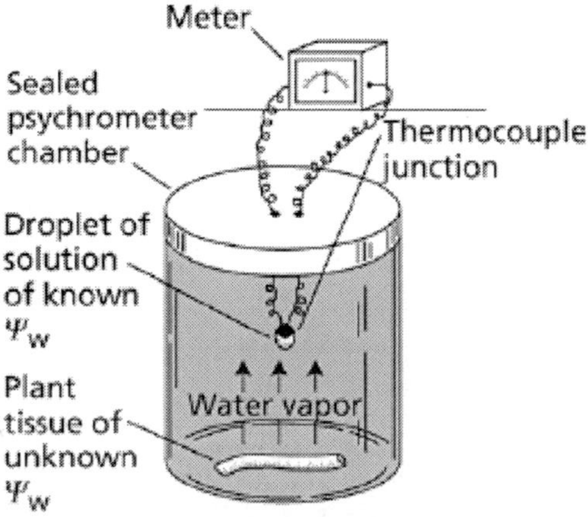

Figure: *Diagram illustrating the use of isopiestic psychrometry to measure the water potential of a plant tissue.*

One psychrometric technique, known as *isopiestic psychrometry*, has been used extensively by John Boyer and coworkers and is illustrated in above Figure. Investigators make a measurement by placing a piece of tissue sealed inside a small chamber that contains a temperature sensor (in this case, a thermocouple) in contact with a small droplet of a standard solution of known solute concentration (known Ψ_s and thus known Ψ_w).

If the tissue has a lower water potential than that of the droplet, water evaporates from the droplet, diffuses through the air, and is absorbed by the tissue. This slight evaporation of water cools the drop.

The larger the difference in water potential between the tissue and the droplet, the higher the rate of water transfer and hence the cooler the droplet. If the standard solution has a lower water potential than that of the sample to be measured, water will diffuse from the tissue to the droplet, causing warming of the droplet.

Measuring the change in temperature of the droplet for several solutions of known Ψ_w makes it possible to calculate the water potential of a solution for which the net movement of water between the droplet and the tissue would be zero signifying that the droplet and the tissue have the same water potential.

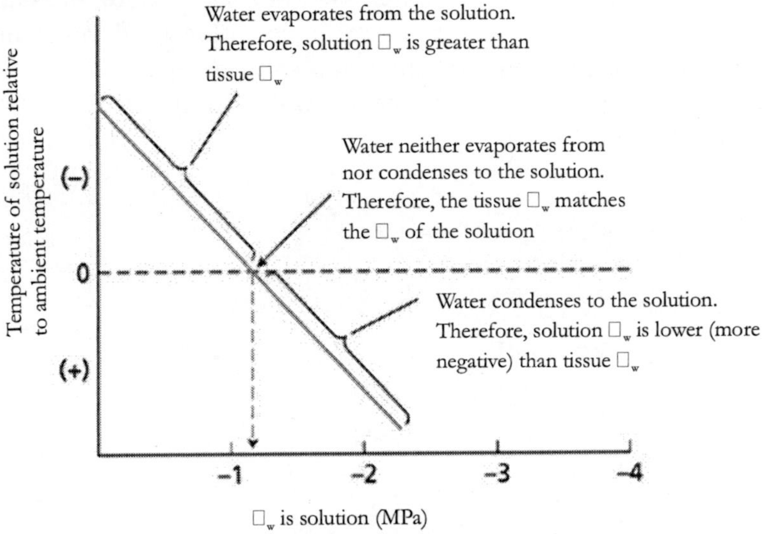

Figure: *The temperature of the sensor in the psychrometer shown in above figure depends on the water potential of the solution relative to the water potential of the tissue sample.*

Psychrometers can be used to measure the water potentials of both excised and intact plant tissue. Moreover, the method can be used to measure the Ψ_s of solutions.

This can be particularly useful with plant tissues. For example, the Ψ_w of a tissue is measured with a psychrometer, and then the tissue is crushed and the Ψ_s value of the expressed cell sap is measured with the same instrument. By combining the two measurements, researchers can estimate the turgor pressure that existed in the cells before the tissue was crushed ($\Psi_p = \Psi_w - \Psi_s$).

A major difficulty with this approach is the extreme sensitivity of the measurement to temperature fluctuations. For example, a change in temperature of 0.01°C corresponds to a change in water potential of about 0.1 MPa. Thus, psychrometers must be operated under constant temperature conditions. For this reason, the method is used primarily in laboratory settings. There are many variations in psychrometric technique; interested readers should consult Brown and Van Haveren 1972, Slavik 1974, and Boyer 1995.

Pressure Chamber

A relatively quick method for estimating the water potential of large pieces of tissues, such as leaves and small shoots, is by use of the pressure chamber. This method was pioneered by Henry Dixon at Trinity College, Dublin, at the beginning of the twentieth century, but it did not come into widespread use until P. Scholander and coworkers at the Scripps Institution of Oceanography improved the instrument design and showed its practical use.

In this technique, the organ to be measured is excised from the plant and is partly sealed in a pressure chamber.

Before excision, the water column in the xylem is under tension. When the water column is broken by excision of the organ (i.e., its tension is relieved allowing its Ψ_p to rise to zero), water is pulled rapidly from the xylem into the surrounding living cells by osmosis.

The cut surface consequently appears dull and dry. To make a measurement, the investigator pressurizes the chamber with compressed gas until the distribution of water between the living cells and the xylem conduits is returned to its initial, pre-excision, state.

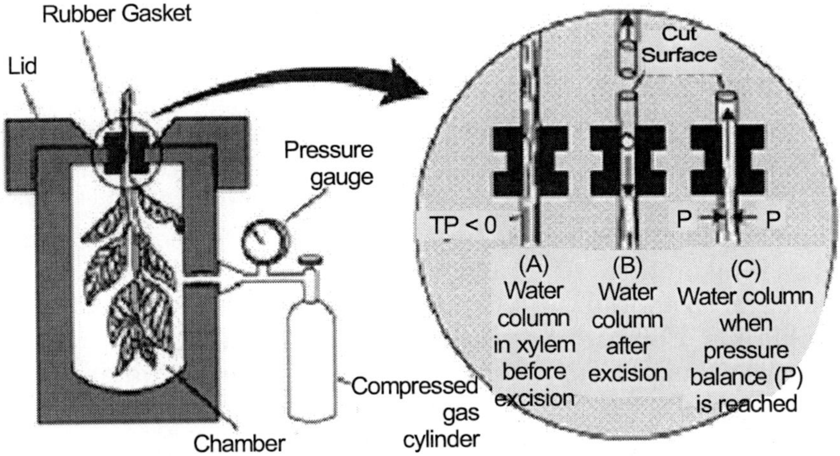

***Figure:** The pressure chamber method for measuring plant water potential. The diagram at left shows a shoot sealed into a chamber, which may be pressurized with compressed gas. The diagrams at right show the state of the water columns within the xylem at three points in time: (A) The xylem is uncut and under a negative pressure, or tension. (B) The shoot is cut, causing the water to pull back into the tissue, away from the cut surface, in response to the tension in the xylem. (C) The chamber is pressurized, bringing the xylem sap back to the cut surface.*

This can be detected visually by observing when the water returns to the open ends of the xylem conduits that can be seen in the cut surface. The pressure needed to bring the water back to its initial distribution is called the *balance pressure* and is readily detected by the change in the appearance of the cut surface, which becomes wet and shiny when this pressure is attained.

The pressure chamber is often described as a tool to measure the tension in the xylem.

However, this is only strictly true for measurements made on a non-transpiring leaf or shoot (for example, one that has been previously enclosed in a plastic bag). When there is no transpiration, the water potential of the leaf cells and the water potential in the xylem will come into equilibrium.

The balancing pressure measured on such a non-transpiring shoot is equal in magnitude but opposite in sign to the pressure in the xylem (Ψ_p). Because the water potential of our non-transpiring leaf is equal to the water potential of the xylem, one can calculate the water potential of the leaf by adding together Ψ_p and Ψ_s of the xylem, provided one collects a sample of xylem sap for determination of Ψ_s.

Luckily Ψ_s of the xylem is usually small (> -0.1 MPa) compared to typical midday tensions in the xylem (Ψ_p of –1 to –2 MPa). Thus, correction for the Ψ_s of the xylem sap is frequently omitted.

Balancing pressure measurements of transpiring leaves are more difficult to interpret. The fact that water is flowing from the xylem to the leaf means that differences in water potential must exist.

When the transpiring leaf or shoot is cut off, the tension in the xylem is instantly relieved and water is drawn into the leaf cells until the water potentials of the xylem and the leaf cells come into equilibrium. Because the total volume of the leaf cells is much larger than the volume of sap in the xylem, this equilibrium water potential will be heavily weighted towards that of the leaf.

Thus, any measurement of the balancing pressure on such a leaf or shoot will result in a value that is approximately the water potential of the leaf, rather than the tension of the xylem. (To be exact, one would have to add the Ψ_s of the xylem sap to the negative of the balancing pressure to get the leaf water potential.) One can explore the differences between the water potential of the xylem and the water potential of a transpiring leaf by comparing balancing pressures measured on covered (i.e., non-transpiring) versus uncovered (transpiring) leaves.

Pressure chamber measurements provide a quick and accurate way of measuring leaf water potential. Because the pressure chamber method does not require delicate instrumentation or temperature control, it has been used extensively under field conditions. For a more complete description of the theory and operation of the pressure chamber.

Cryoscopic Osmometer

The cryoscopic osmometer measures the osmotic potential of a solution by measuring its freezing point.

Solutions have *colligative* properties that collectively depend on the number of dissolved particles and not on the nature of the solute. For example, solutes reduce the vapour pressure of a solution, raise its boiling point, and lower its freezing point.

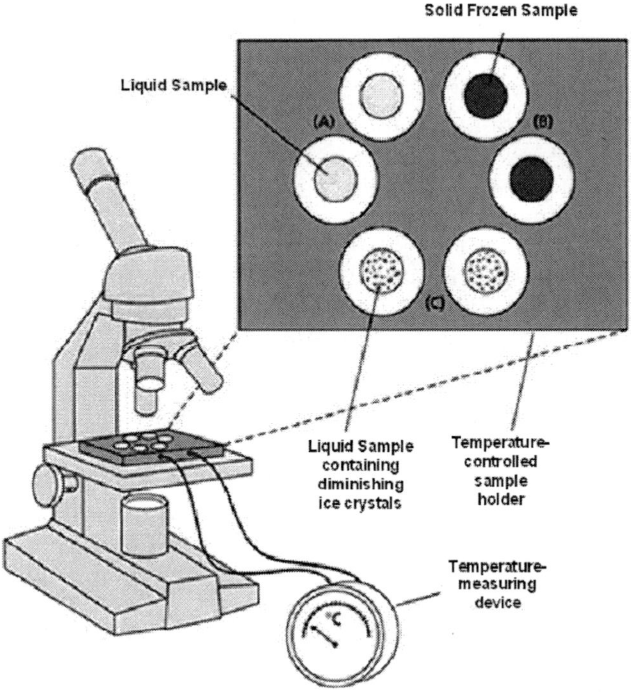

Figure: *A cryoscopic osmometer measures the concentration of total dissolved solutes by measuring the freezing-point depression of a solution. (A) Very small liquid samples are loaded onto the temperature-controlled stage of a microscope. (B) When the temperature is quickly reduced, the samples supercool and freeze. (C) Slowly warming the stage causes the samples to thaw. The temperature at which the last ice crystal melts provides a measure of the melting point of the sample.*

The specific nature of the solute does not matter. One of the colligative *properties* of solutions is the decrease in the freezing point as the solute concentration increases. For example, a solution containing 1 mol of solutes per kilogram of water has a freezing point of –1.86°C, compared with 0°C for pure water.

Various instruments can be used to measure the freezing-point depression of solutions. With a cryoscopic osmometer, solution samples as small as 1 nanoliter (10^{-9} L) are placed in an oil medium located on the temperature-controlled stage of a microscope. The very small sample size allows sap from single cells to be measured and permits rapid thermal equilibration with the stage.

To prevent evaporation, the investigator suspends the samples in oil-filled wells in a silver plate (silver has high thermal conductivity).

The temperature of the stage is rapidly decreased to about –30° C, which causes the sample to freeze. The temperature is then raised very slowly, and the melting process in the sample is observed through the microscope. When the last ice crystal in the sample melts, the temperature of the stage is recorded (note that the melting and freezing points are the same).

It is straightforward to calculate the solute concentration from the freezing-point depression; and from the solute concentration (c_s), Ψ_s is calculated as $-RTc_s$. This technique has been used to measure droplets extracted from single cells.

Pressure Probe

If a cell were as large as a watermelon or even a grape, measuring its hydrostatic pressure would be a relatively easy task. Because of the small size of plant cells, however, the development of methods for direct measurement of turgor pressure has been slow.

Using a micromanometer, Paul Green at the University of Pennsylvania developed one of the first direct methods for measuring turgor pressure in plant cells (Green and Stanton 1967). In this technique, an air-filled glass tube sealed at one end is inserted into a cell.

The high pressure in the cell compresses the trapped gas, and from the change in volume one can readily calculate the pressure of the cell from the ideal gas law (pressure × volume = constant). This method works only for cells of relatively large volume, such as the giant cell of the filamentous green alga *Nitella*. For smaller cells, the loss of cell sap into the glass tube is sufficient to deflate the cell and this yields artifactually low pressures. For higher plant cells, which are several orders of magnitude smaller in volume than *Nitella,* a more sophisticated device, the pressure probe, was developed by Ernest Steudle, Ulrich Zimmermann, and their colleagues in Germany. This instrument is similar to a miniature syringe.

A glass microcapillary tube is pulled to a fine point and is inserted into a cell. The microcapillary is filled with silicone oil, a relatively incompressible fluid that can be readily distinguished from cell sap under a microscope.

When the tip of the microcapillary is first inserted into the cell, cell sap begins to flow into the capillary because of the initial low pressure of that region. Investigators can observe such movement of sap under the microscope and counteract it by pushing on the plunger of the device, thus building up a pressure. In such fashion the boundary between the oil and the cell sap can be pushed back to the tip of the microcapillary.

When the boundary is returned to the tip and is held in a constant position, the initial volume of the cell is restored and the pressure inside the cell is exactly balanced by the pressure in the capillary. This pressure is measured by a pressure sensor in the device. Thus the hydrostatic pressure of individual cells may be measured directly.

This method has been used to measure Ψ_p and other parameters of water relations in cells of both excised and intact tissues of a variety of plant species. The primary limitation of this method is that some cells are too small to measure.

Furthermore, some cells tend to leak after being stabbed with the capillary, and others plug up the tip of the capillary, thereby preventing valid measurements. The pressure probe has also been adapted to measure positive and negative values of Ψ_p in the xylem. However, technical problems with cavitation limit the measurement of negative Ψ_p by this technique.

Understanding Hydraulic Conductivity

Consider a cell with an initial water potential of –0.2 MPa, submerged in pure water. From this information we know that water will flow into the cell and that the driving force is $\Delta\Psi_w = 0.2$ MPa, but what is the initial rate of movement? The rate depends on permeability of the membrane to water, a property usually called the hydraulic conductivity (*Lp*) of the membrane.

Driving force, membrane permeability, and flow rate are related by the following equation:

$$\text{Flow rate} = \text{driving force} \times \text{hydraulic conductivity}$$

Hydraulic conductivity expresses how readily water can move across a membrane and has units of volume of water per unit area of membrane per unit time per unit driving force (for instance, $m^3\ m^{-2}\ s^{-1}\ MPa^{-1}$ or $m\ s^{-1}\ MPa^{-1}$). The larger the hydraulic conductivity, the larger the flow rate. The hydraulic conductivity of the membrane is $10^{-6}\ m\ s^{-1}\ MPa^{-1}$. The transport (flow) rate (J_v) can then be calculated from the following equation:

$$J_v = (Lp(\Delta\psi_w)$$

where J_v is the volume of water crossing the membrane per unit area of membrane and per unit time ($m^3\ m^{-2}\ s^{-1}$ or, equivalently, $m\ s^{-1}$). Please note that this equation assumes that the membrane is ideal—that is, that solute transport is negligible and water transport is equally sensitive to $\Delta\Psi_s$ and $\Delta\Psi_p$ across the membrane. Nonideal membranes require a more complicated equation that separately accounts for water flow induced by $\Delta\Psi_s$ and by $\Delta\Psi_p$.

In our example, J_v has a value of $0.2 \times 10^{-6}\ m\ s^{-1}$. Note that J_v has the physical meaning of a velocity. We can calculate the flow rate in volumetric terms ($m^3\ s^{-1}$) by multiplying J_v by the surface area of the cell.

The resulting value is the *initial* rate of water transport. As water is taken up, cell Ψ_w increases and the driving force ($\Delta\Psi_w$) decreases. As a result, water transport slows with time. As elaborated in p. 43 of the textbook, the rate approaches zero in an exponential manner.

Wilting and Plasmolysis

Plasmolysis is the separation of plant cell cytoplasm from the cell wall as a result of water loss. It is unlikely to occur in nature, except in severe conditions. Plasmolysis is induced in the laboratory by immersing a plant cell in a strongly saline or sugary solution, so that water is lost by osmosis.

If onion epidermal tissue is immersed in a solution of calcium nitrate, cells rapidly lose water by osmosis and the protoplasm of the cells shrinks. This occurs because the calcium and nitrate ions freely permeate the cell wall and encounter the selectively permeable plasma membrane.

The large vacuole in the centre of the cell originally contains a dilute solution with much lower osmotic pressure than that of the calcium nitrate solution on the other side of the membrane. The vacuole thus loses water and becomes smaller. The space between the cell membrane and the cell wall enlarges and the plasma membrane and the protoplasm within it contract to the centre of the cell. Strands of cytoplasm extend to the cell wall because of plasma membrane-cell wall attachment points. Plasmolysed cells die unless they are transferred quickly from the salt or sugar solution to water.

Mineral Nutrition

Symptoms of Deficiency in Essential Minerals

Visual nutrient deficiency symptoms can be a very powerful diagnostic tool for evaluating the nutrient status of plants. One should keep in mind, however, that a given individual visual symptom is seldom sufficient to make a definitive diagnosis of a plant's nutrient status.

Many of the classic deficiency symptoms such as tip burn, chlorosis and necrosis are characteristically associated with more than one mineral deficiency and also with other stresses that by themselves are not diagnostic for any specific nutrient stress. However, their detection is extremely useful in making an evaluation of nutrient status.

In addition to the actual observations of morphological and spectral symptoms, knowing the location and timing of these symptoms is a critical aspect of any nutrient status evaluation. Plants do not grow in isolation, they are part of the overall environment and as such they respond to environmental changes as that affect nutrient availability. Also, plants do influence their environment and can contribute to environmental changes, which in turn can affect the nutrient status of the plant.

Sources of Visual Symptoms

Stresses such as salinity, pathogens, and air pollution induce their own characteristic set of visual symptoms. Often, these symptoms closely resemble those of nutrient deficiency.

Pathogens often produce an interveinal chlorosis, and air pollution and salinity stress can cause tip burn. Although at first these symptoms might seem similar in their general appearance to nutrient deficiency symptoms, they do differ in detail and/or in their overall developmental pattern.

Pathological symptoms can often be separated from nutritional symptoms by their distribution in a population of affected plants. If the plants are under a nutrient stress, all plants of a given type and age in the same environment tend to develop similar symptoms at the same time.

However if the stress is the result of pathology, the development of symptoms will have a tendency to vary between plants until a relatively advanced stage of the pathology is reached.

Environmental Associations

Plants remove substantial amounts of nutrients from the soil during their normal growth cycle and many long-term environmental changes occur as a result of this process. Effects on the soil go considerably beyond the straight removal or depletion of nutrients. Charge balance must be maintained in the plant-soil system during nutrient uptake. Charge balance is usually achieved by the excretion of proton and/or hydroxyl ions by the plant to replace the absorbed nutrient cations or anions.

For example when plants are fertilized with ammonia, they acquire most of their nitrogen in the form of the ammonium cation, rather than from the usual nitrate anion. Because nitrate is the only anion used by the plant in large amounts, the net result of this change is that during normal nutrient uptake the proton excretion will far exceed that of hydroxyl ions.

In the case of vigorously growing plants, the amount of excreted protons can be sufficiently large as to decrease the pH of the soil by several pH units. Changes in soil pH of such magnitude can have large implications for a number of soil processes such as soil structure, nutrient availability and leaching of nutrients.

The immediate effect on the soil may be favourable for some plants, especially acid-loving plants, in that it tends to make iron more available. However, in the long run, lowering the soil pH can be deleterious to plants in that the availability of nutrients will change. A lower soil pH will allow micronutrients to be more readily leached from the soil profile, eventually resulting in deficiencies of nutrients such as Cu and Zn. Additionally, when the pH of the soil drops much below pH 5, the solubility of Al and Mn can increase to such an extent as to become toxic to most plant growth.

Plants are often thought of as passive in relation to the environment. However this is not always a valid assumption; for there are many plants that clearly manipulate their environment in a fashion that tends to makes certain nutrients more readily available. For example, iron is a limiting nutrient in many agricultural areas, but it comprises about 3% of the average soil which, if available, would be far in excess of the needs of the average plant. Some plants actively excrete protons, and the resulting decrease in pH increases the solubility of iron in their environment. In addition, other plants excrete phytosiderophores that chelate the soil iron rendering it a more available form for the plants.

Pathways of Symptom Development

At first glance, it would appear that the distinction of deficiency symptoms for the 13 known essential mineral nutrients should be relatively simple. But such an assumption is incorrect. In fact, the deficiency symptoms are quite complex because each nutrient has a number of different biological functions and each function may have an independent set of interactions with a wide range of environmental parameters.

In addition, the expression of these symptoms varies for acute or chronic deficiency conditions. Acute deficiency occurs when a nutrient is suddenly no longer available to a rapidly growing plant. Chronic deficiency occurs when there is a limited but continuous supply of a nutrient, at a rate that is insufficient to meet the growth demands of the plant. Most of the classic deficiency symptoms described in textbooks are characteristic of acute deficiencies. The most common symptoms of low-grade, chronic deficiencies are a tendency towards darker green leaves and stunted or slow growth.

Typically most published descriptions of deficiency symptoms arise from experiments conducted in greenhouses or growth chambers where the plants are grown in hydroponics or in media where the nutrients are fully available. In these conditions, nutrients are readily available while present, but when a nutrient is depleted, the plant suddenly faces an acute deficiency. Thus, hydroponic studies favour the development of acute deficiencies.

In experiments designed to study micronutrient deficiency symptoms, micronutrients are usually omitted from the nutrient solution. Micronutrients are often present in the seed or as contaminants in the environment, so a plant of adequate size will exhaust these trace amounts of micronutrient and develop characteristic acute deficiency systems.

When deficiency symptoms of macronutrients are sought, the macronutrient is removed suddenly from a suitable sized rapidly growing plant. Alternatively the plant can initially be given a one-time supply of the nutrient that is sufficient for a limited amount of growth. Because macronutrients are continuously required in relatively large amounts by rapidly growing plants, the available nutrients will be rapidly depleted, resulting in an acute deficiency.

In natural systems, the plant encounters many degrees and types of stresses that result in different types of symptoms occurring over time. Perhaps the most common nutrient deficiency in natural environments is the case of a limited nutrient supply that is continuously renewed at a low rate from soil weathering processes. In such cases, the limited nutrient availability results in chronic nutrient deficiency symptoms.

Effect of Nutrient Mobility on Symptom Development

The interaction between nutrient mobility in the plant, and plant growth rate can be a major factor influencing the type and location of deficiency symptoms that develop.

For very mobile nutrients such as nitrogen and potassium, deficiency symptoms develop predominantly in the older and mature leaves. This is a result of these nutrients being preferentially mobilized during times of nutrient stress from the older leaves to the newer leaves near the growing regions of the plant. Additionally, mobile nutrients newly acquired by the roots are also preferentially translocated to new leaves and the growing regions.

Thus old and mature leaves are depleted of mobile nutrients during times of stress while the new leaves are maintained at a more favourable nutrient status.

The typical localization of deficiency symptoms of very weakly mobile nutrients such as calcium, boron, and iron is the opposite to that of the mobile nutrients; these deficiency symptoms are first displayed in the growing regions and new leaves while the old leaves remain in a favourable nutrient status. (This assumes that these plants started with sufficient nutrient, but ran out of nutrient as they developed).

In plants growing very slowly for reasons other than nutrition (such as low light) a normally limiting supply of a nutrient could, under these conditions, be sufficient for the plant to slowly develop, maybe even without symptoms. This type of development is likely to occur in the case of weakly mobile nutrients because excess nutrients in the older leaves will eventually be mobilized to supply newly developing tissues.

In contrast, a plant with a similar supply that is growing rapidly will develop severe deficiencies in the actively growing tissue such as leaf edges and the growing region of the plant. A classic example of this is calcium deficiency in vegetables such as lettuce where symptoms develop on the leaf margins (tip burn) and the growing region near the meristems.

The maximal growth rate of lettuce is often limited by the internal translocation rate of calcium to the growing tissue rather than from a limited nutrient supply in the soil.

When moderately mobile nutrients such as sulfur and magnesium are the limiting nutrients of the system, deficiency symptoms are normally seen over the entire plant. However the growth rate and rate of nutrient availability can make a considerable difference on the locations at which the symptoms develop. If the nutrient supply is marginal compared to the growth rate, symptoms will appear on the older tissue, but if the nutrient supply is very low compared to the growth rate, or the nutrient is totally depleted, the younger tissue will become deficient first.

Plant Competition and Induced Deficiencies

When the observed symptoms are the direct result of a nutrient deficiency, the actions needed for correction are relatively straightforward. However symptoms are often the result of interactions with other environmental factors limiting the availability of the nutrient whose symptoms are expressed. The classic instance is that of iron deficiency induced by an excess of heavy metals in the environment. Transition metals such as Cu, Zn Cr and Ni compete with Fe and each other for plant uptake. Competition for uptake is not specific to Fe and heavy metals but is true for all mineral nutrients that are chemically similar and have similar uptake mechanisms. For example if the availability of Cu or Zn is relatively less than that of Fe, then excessive concentrations of some other metal such as Ni or Cr will induce a deficiency of one of these nutrients rather than Fe. In the case of the macronutrients, excessive amounts of Mg will compete with K for uptake and can possibly induce a K deficiency. The barrenness of serpentine soils is the result of such competition, with the high Mg of these soils inducing a Ca deficiency. The toxicity of a low pH soil is another example of a basic nutrient deficiency. Low pH has a two-fold effect on soil nutrients: It enhances the leaching of cations, reducing their availability in the soil, and the relatively abundant protons in the soil compete with Ca and other cations for uptake. Thus, nutrient deficiencies can be induced by a number of different mechanisms often working in concert to limit the availability of a nutrient.

Nutrient Demand and Use Efficiency

Although all plants of the same species respond similarly to nutrient stress, plants of similar species will often show significant differences in their nutrient use efficiency. This results from differences in growth rate, root distribution, phase of development, and efficiency of nutrient uptake and utilization. This implies that in any given location, plants from one species may become nutrient-deficient, while those from another species growing in the same environment right next to them, may not show any deficiency symptoms.

Growth rate also affects nutrient status. When the nutrient supply is barely inadequate for growth under existing environmental conditions, many plants adjust their growth rate to match that supported by the available nutrient supply without displaying typical visual deficiency symptoms. Agricultural systems differ from natural systems in that crop plants have been selected primarily for rapid growth under low stress conditions.

This rapid growth rate results in a high nutrient demand by these plants and a higher incidence of nutrient deficiency unless supplemental fertilizers are supplied. It is not uncommon to find agricultural crops showing severe signs of nutrient stress, with native plants growing in the same area showing little or no indication of nutrient stress. In agriculture systems chronic deficiency symptoms develop mostly in crops with little or limited fertilization. Acute nutrient deficiency symptoms most often occur when new crops with a higher nutrient demand are introduced, or less productive lands are brought under cultivation for the production of rapidly growing crop plants.

Uniformity of Nutrient Status

Not all tissues of a plant are at the same nutrient status during times of stress. Leaves on the same plant that are exposed to different environmental conditions, (such as light), or those of different ages may have considerable differences in nutrient status. Mineral nutrients are for the most part acquired by the roots and translocated throughout the plant. The distance of any part of the plant to the roots will influence nutrient availability, particularly in the case of the less mobile nutrients. In plants recovering from nutrient deficiency, the root and conductive tissues recover first. For example, in the case of recovery from Fe deficiency, it is common to see the veins re-green while the interveinal tissue remains chlorotic and Fe-deficient.

In order to maintain rapid, optimal growth, all plant tissues must have a favourable nutrient status. Although a plant may be marginally low in a number of nutrients, only one nutrient at a time will limit overall growth. However, if the supply of that limiting nutrient is increased even slightly, the resulting increase in growth will increase the demand for all other nutrients and another nutrient, the next lowest in availability, will become limiting.

Other Diagnostic Tools

Although visual diagnostic symptoms are an extremely valuable tool for the rapid evaluation of the nutrient status of a plant, they are only some of the tools available. Other major tools include microscopic studies, spectral analysis, and tissue and soil analysis. These methods all vary in their precision, rapidity and their ability to predict future nutrient status. Because of the close interaction between plant growth and the environment, all predictions of future nutrient status must make assumptions about how the environment will change in that time frame.

The principle advantage of visual diagnostic symptoms is that they are readily obtained and provide an immediate evaluation of nutrient status. Their main drawback is that the visual symptoms do not develop until after there has been a major effect on yield, growth and development.

Tissue analysis is nutrient-specific but relatively slow; tissues must be sampled, processed and analysed before the nutrient status can be determined. An analysis of the mineral nutrient content of selected plants tissues, when compared against Critical Level values, can be used to evaluate the plant nutrient status at the time of sampling with a relatively high degree of confidence and can be extrapolated to project nutrient status at harvest. Soil analysis is similar to tissue analysis but evaluates the potential supplying power of the soil instead of plant nutrient status. Plant analysis provides information as to what the plant needs, while soil analysis provides information about the status of the nutrient supply.

Spectral analysis of nutrient status is still in its infancy and is presently used primarily in the inventory of global resources and in specialized studies. Microscopic studies are most valuable in looking at the physiological aspects of nutrient stress rather than the evaluation of plant nutrient status on a whole plant or crop basis.

Symptom Descriptions

It is unusual to find any one leaf or even one plant that displays the full array of symptoms that are characteristic of a given deficiency. It is thus highly desirable to know how individual symptoms look, for it is possible for them to occur in many possible combinations on a single plant. Most of the terms used below in the description of deficiency symptoms are reasonably self evident; a few however have a distinct meaning in the nutrient deficiency field. For example, the term *chlorotic,* which is a general term for yellowing of leaves through the loss of chlorophyll, cannot be used without further qualification because there may be an overall chlorosis as in nitrogen deficiency, interveinal, as in iron deficiency, or marginal, as in calcium deficiency. Another term used frequently in the description of deficiency symptoms is *necrotic,* a general term for brown, dead tissue. This symptom can also appear in many varied forms, as is the case with chlorotic symptoms.

Nutrient deficiency symptoms for many plants are similar, but because of the large diversity found in plants and their environments there is a range of expression of symptoms. Because of their parallel veins, grasses and other monocots generally display the affects of chlorosis as a series of stripes rather than the netted interveinal chlorosis commonly found in dicots. The other major difference is that the marginal necrosis or chlorosis found in dicots is often expressed as tip burn in monocots.

Magnesium. The Mg-deficient leaves show advanced interveinal chlorosis, with necrosis developing in the highly chlorotic tissue. In its advanced form, magnesium deficiency may superficially resemble potassium deficiency. In the case of magnesium deficiency the symptoms generally start with mottled chlorotic areas developing in the interveinal tissue. The interveinal laminae tissue tends to expand proportionately more than the other leaf tissues, producing a raised puckered surface, with the top of the puckers progressively going from chlorotic to necrotic tissue. In some plants such as the Brassica (The mustard family, which includes vegetables such as broccoli, brussel sprouts, cabbage, cauliflower, collards, kale, kohlrabi, mustard, rape, rutabaga and turnip.), tints of orange, yellow, and purple may also develop.

Figure: *Magnesium deficiency symptoms in tomato.*

Manganese. These leaves show a light interveinal chlorosis developed under a limited supply of Mn. The early stages of the chlorosis induced by manganese deficiency are somewhat similar to iron deficiency. They begin with a light chlorosis of the young leaves and netted veins of the mature leaves especially when they are viewed through transmitted light. As the stress increases, the leaves take on a gray metallic sheen and develop dark freckled and necrotic areas along the veins.

Figure: *Molybdenum deficiency symptoms in tomato.*

A purplish luster may also develop on the upper surface of the leaves. Grains such as oats, wheat, and barley are extremely susceptible to manganese deficiency. They develop a light chlorosis along with gray specks which elongate and coalesce, and eventually the entire leaf withers and dies.

Molybdenum. These leaves show some mottled spotting along with some interveinal chlorosis. An early symptom for molybdenum deficiency is a general overall chlorosis, similar to the symptom for nitrogen deficiency but generally without the reddish coloration on the undersides of the leaves. This results from the requirement for molybdenum in the reduction of nitrate, which needs to be reduced prior to its assimilation by the plant.

Figure: *Molybdenum deficiency symptoms in tomato.*

Thus, the initial symptoms of molybdenum deficiency are in fact those of nitrogen deficiency. However, molybdenum has other metabolic functions within the plant, and hence there are deficiency symptoms even when reduced nitrogen is available. In the case of cauliflower, the lamina of the new leaves fail to develop, resulting in a characteristic whiptail appearance. In many plants there is an upward cupping of the leaves and mottled spots developing into large interveinal chlorotic areas under severe deficiency. At high concentrations, molybdenum has a very distinctive toxicity symptom in that the leaves turn a very brilliant orange.

Nitrogen. The chlorotic symptoms shown by this leaf resulted from nitrogen deficiency. A light red cast can also be seen on the veins and petioles. Under nitrogen deficiency, the older mature leaves gradually change from their normal characteristic green appearance to a much paler green. As the deficiency progresses these older leaves become uniformly yellow (chlorotic). Leaves approach a yellowish white colour under extreme deficiency.

The young leaves at the top of the plant maintain a green but paler colour and tend to become smaller in size. Branching is reduced in nitrogen deficient plants resulting in short, spindly plants. The yellowing in nitrogen deficiency is uniform over the entire leaf including the veins. However in some instances, an interveinal necrosis replaces the chlorosis commonly found in many plants. In some plants the underside of the leaves and/or the petioles and midribs develop traces of a reddish or purple colour. In some plants this coloration can be quite bright. As the deficiency progresses, the older leaves also show more of a tendency to wilt under mild water stress and become senescent much earlier than usual. Recovery of deficient plants to applied nitrogen is immediate (days) and spectacular.

Figure: *Nitrogen deficiency symptoms in tomato.*

Phosphorus. These phosphorus-deficient leaves show some necrotic spots. As a rule, phosphorus deficiency symptoms are not very distinct and thus difficult to identify. A major visual symptom is that the plants are dwarfed or stunted. Phosphorus deficient plants develop very slowly in relation to other plants growing under similar environmental conditions but without phosphorus deficiency. Phosphorus deficient plants are often mistaken for unstressed but much younger plants. Some species such as tomato, lettuce, corn and the brassicas develop a distinct purpling of the stem, petiole and the under sides of the leaves. Under severe deficiency conditions there is also a tendency for leaves to develop a blue-gray luster. In older leaves under very severe deficiency conditions a brown netted veining of the leaves may develop.

Figure: *Phosphorus deficiency symptoms in tomato.*

Sulfur. This leaf shows a general overall chlorosis while still retaining some green colour. The veins and petioles show a very distinct reddish colour. The visual symptoms of sulfur deficiency are very similar to the chlorosis found in nitrogen deficiency. However, in sulfur deficiency the yellowing is much more uniform over the entire plant including young leaves.

Figure: *Sulfur deficiency symptoms in tomato.*

The reddish colour often found on the underside of the leaves and the petioles has a more pinkish tone and is much less vivid than that found in nitrogen deficiency. With advanced sulfur deficiency brown lesions and/or necrotic spots often develop along the petiole, and the leaves tend to become more erect and often twisted and brittle.

Zinc. This leaf shows an advanced case of interveinal necrosis. In the early stages of zinc deficiency the younger leaves become yellow and pitting develops in the interveinal upper surfaces of the mature leaves. Guttation is also prevalent. As the deficiency progress these symptoms develop into an intense interveinal necrosis but the main veins remain green, as in the symptoms of recovering iron deficiency. In many plants, especially trees, the leaves become very small and the internodes shorten, producing a rosette like appearance.

Figure: *Zinc deficiency symptoms in tomato.*

Boron. These boron-deficient leaves show a light general chlorosis. The tolerance of plants to boron varies greatly, to the extent that the boron concentrations necessary for the growth of plants having a high boron requirement may be toxic to plants sensitive to boron. Boron is poorly transported in the phloem of most plants, with the exception of those plants that utilize complex sugars, such as sorbitol, as transport metabolites. In a recent study, tobacco plants engineered to synthesize sorbitol were shown to have increased boron mobility, and to better tolerate boron deficiency in the soil.

In plants with poor boron mobility, boron deficiency results in necrosis of meristematic tissues in the growing region, leading to loss of apical dominance and the development of a rosette condition. These deficiency symptoms are similar to those caused by calcium deficiency. In plants in which boron is readily transported in the phloem, the deficiency symptoms localize in the mature tissues, similar to those of nitrogen and potassium. Both the pith and the epidermis of stems may be affected, often resulting in hollow or roughened stems along with necrotic spots on the fruit.

The leaf blades develop a pronounced crinkling and there is a darkening and crackling of the petioles often with exudation of syrupy material from the leaf blade. The leaves are unusually brittle and tend to break easily. Also, there is often a wilting of the younger leaves even under an adequate water supply, pointing to a disruption of water transport caused by boron deficiency.

Calcium. These calcium-deficient leaves show necrosis around the base of the leaves. The very low mobility of calcium is a major factor determining the expression of calcium deficiency symptoms in plants. Classic symptoms of calcium deficiency include blossom-end rot of tomato (burning of the end part of tomato fruits), tip burn of lettuce, blackheart of celery and death of the growing regions in many plants.

All these symptoms show soft dead necrotic tissue at rapidly growing areas, which is generally related to poor translocation of calcium to the tissue rather than a low external supply of calcium. Very slow growing plants with a deficient supply of calcium may re-translocate sufficient calcium from older leaves to maintain growth with only a marginal chlorosis of the leaves.

This ultimately results in the margins of the leaves growing more slowly than the rest of the leaf, causing the leaf to cup downward. This symptom often progresses to the point where the petioles develop but the leaves do not, leaving only a dark bit of necrotic tissue at the top of each petiole. Plants under chronic calcium deficiency have a much greater tendency to wilt than non-stressed plants.

Chloride. These leaves have abnormal shapes, with distinct interveinal chlorosis. Plants require relatively high chlorine concentration in their tissues. Chlorine is very abundant in soils, and reaches high concentrations in saline areas, but it can be deficient in highly leached inland areas.

The most common symptoms of chlorine deficiency are chlorosis and wilting of the young leaves. The chlorosis occurs on smooth flat depressions in the interveinal area of the leaf blade. In more advanced cases there often appears a characteristic bronzing on the upper side of the mature leaves. Plants are generally tolerant of chloride, but some species such as avocados, stone fruits, and grapevines are sensitive to chlorine and can show toxicity even at low chloride concentrations in the soil.

Copper. These copper-deficient leaves are curled, and their petioles bend downward. Copper deficiency may be expressed as a light overall chlorosis along with the permanent loss of turgor in the young leaves. Recently matured leaves show netted, green veining with areas bleaching to a whitish gray. Some leaves develop sunken necrotic spots and have a tendency to bend downward. Trees under chronic copper deficiency develop a rosette form of growth. Leaves are small and chlorotic with spotty necrosis.

Iron. These iron-deficient leaves show strong chlorosis at the base of the leaves with some green netting. The most common symptom for iron deficiency starts out as an interveinal chlorosis of the youngest leaves, evolves into an overall chlorosis, and ends as a totally bleached leaf. The bleached areas often develop necrotic spots. Up until the time the leaves become almost completely white they will recover upon application of iron. In the recovery phase the veins are the first to recover as indicated by their bright green colour.

This distinct venial re-greening observed during iron recovery is probably the most recognizable symptom in all of classical plant nutrition. Because iron has a low mobility, iron deficiency symptoms appear first on the youngest leaves. Iron deficiency is strongly associated with calcareous soils and anaerobic conditions, and it is often induced by an excess of heavy metals.

Potassium. Some of these leaves show marginal necrosis (tip burn), others at a more advanced deficiency status show necrosis in the interveinal spaces between the main veins along with interveinal chlorosis. This group of symptoms is very characteristic of K deficiency symptoms. The onset of potassium deficiency is generally characterized by a marginal chlorosis progressing into a dry leathery tan scorch on recently matured leaves.

This is followed by increasing interveinal scorching and/or necrosis progressing from the leaf edge to the midrib as the stress increases.

As the deficiency progresses, most of the interveinal area becomes necrotic, the veins remain green and the leaves tend to curl and crinkle. In some plant such as legumes and potato, the initial symptom of deficiency is white speckling or freckling of the leaf blades.

In contrast to nitrogen deficiency, chlorosis is irreversible in potassium deficiency, even if potassium is given to the plants. Because potassium is very mobile within the plant, symptoms only develop on young leaves in the case of extreme deficiency.

Potassium deficiency can be greatly alleviated in the presence of sodium but the resulting sodium-rich plants are much more succulent than a high potassium plant. In some plants over 90% of the required potassium can be replaced with sodium without any reduction in growth.

Bibliography

Akhil Baruah: *Advanced Morphology of Angiosperms*, Aavishkar, Delhi, 2008.

Alfred Steferud: *Diseases of Fruits and Nuts*, Biotech Books, Delhi, 2005.

Andrews J.: *The Domesticated Capsicums*, University of Texas Press. 1995.

Arthur W. Gilbert, Mortier F. Barrus and Daniel Dean: *Growing and Breeding of Potatoes*, Asiatic Pub, Delhi, 2006.

Bahar A. Siddiqui and Samiullah Khan: *Plant Breeding Advances and in vitro Culture*, CBS, Delhi, 1997.

Barbeau G.: *Tropical Fruits in Nicaragua*, Managua, Nicaragua Ministerio de Desarrollo Agropecuario, Agraria, 1990.

Chadha K. L. and Pareek O. P.: *Advances in Horticulture: Fruit Crops,* New Delhi, Malhotra Publishing House, 1993.

Cronquist, A.: *The Evolution and Classification of Flowering Plants*, New York Botanical Garden, Bronx, New York., 1988

Crucefix D.: *Avocado Variety Selection for Export Development,* Roseau, CARDI, 1996.

Daniells J.: *Illustrated Guide to the Identification of Banana varieties in the South Pacific*, Canberra, ACIAR, 1995.

David White: *The Physiology and Biochemistry of Prokaryotes*, Oxford University Press, Delhi, 2007.

Dinabandhu Sahoo: *Farming the Ocean : Seaweeds Cultivation and Utilization*, Aravali, Delhi, 2000.

Featherly H. I.: *Taxonomic Terminology of the Higher Plants*, USA, Iowa State College Press, 1954.

Ferentinos L.: *Proceeding of the Sustainable Taro Culture for the Pacific Conference*, Honolulu, HITAHR, 1993.

Glowinski Louis: *The Complete Book of Fruit Growing in Australia*, Australia, Lothian Publishing Company Pty. Ltd., 1991.

Herbert Meschkowski: *Ways of Thought of Great Mathematicians*, Holden-Day Inc., San Francisco, 1964.

Huaman Z.: *Descriptors for Sweet Potato*, Rome, International Board for Plant Genetic Resources, 1991.

Jacquat Christiane: *Plants from the Markets of Thailand*, Bangkok, Duang Kamol, 1990.

Keshav Prasad Yadav: *Application of Morphometry in Geomorphology*, Radha Pub, Delhi, 2008.

Lorenzen, S.: *The Phylogenetic Systematics of Freeliving Nematodes*, London, The Ray Society, 1994.

Madulid Domingo A.: *A Pictorial Cyclopedia of Philippine Ornamental Plants*, Philippines, Makati Metro Manila, 1995.

Nobel, P. S.: *Physicochemical and Environmental Plant Physiology*, Academic Press, San Diego, 1999.

Oldham P.: *Cost of Production of Major Tree Crops in Dominica*, Roseau, Ministry of Agriculture, 1991.

Pemberton, R. W.: *Predictable Risk to Native Plants in Weed Biological Control,* Oecologia, 2000.

Pramod P Mahulikar and Kshama M Chavan: *Botanicals as Ecofriendly Pesticides*, New India Pub Agency, Delhi, 2007.

Qystein V. Sjaastad: *Physiology of Domestic Animals*, International Book Distributing Co., Delhi, 2005.

Ragone D.: *Breadfruit: Artocarpus Altilis (Parkinson) Fosberg*, Rome, International Plant Genetic Resources Institute, 1997.

Sanchez, P. A.: *Properties and Management of Soils in the Tropics*, John Wiley and Sons, New York, 1976.

Siddiqi, M.R.: *Tylenchida Parasites of Plants and Insects,* Wallingford, UK, CAB International, 1986.

Sinha G. C., Reddy Y. T. N. and Singh G.: *Propagation and Rootstocks in Guava*, New Delhi, Malhotra Publishers, 1993.

Tawde A. B.: *Propagation and Rootstocks of Mango*, New Delhi, Malhotra, 1993.

Urton, Gary: *The Social Life of Numbers*, Austin, University of Texas Press, 1997.

Vanderplank, J.E.: *Plant Diseases: Epidemics and Control,* New York, NY, USA, Academic Press, 1963.

Vikas Chaudhary: *Entomology and Pest Management*, Navyug, Delhi, 2008.

Woolfe Jennifer A.: *The Potato in the Human Diet*, Cambridge, Cambridge University Press, 1989.

Yoshida T.: *Cultivation of Citrus Genetic Resources for Evaluation of Characteristics*, Tokyo, JICA, 1996.

Index

R

S

T

V

W

❑❑❑